Spectacular Science, Technology and Superstition in the Age of Shakespeare

Edited by Sophie Chiari and Mickaël Popelard

EDINBURGH
University Press

Edinburgh University Press is one of the leading university presses in the UK. We publish academic books and journals in our selected subject areas across the humanities and social sciences, combining cutting-edge scholarship with high editorial and production values to produce academic works of lasting importance. For more information visit our website: edinburghuniversitypress.com

Edinburgh University Press Ltd
The Tun – Holyrood Road, 12(2f) Jackson's Entry, Edinburgh EH8 8PJ

Typeset in 11/13 Adobe Sabon by
IDSUK (DataConnection) Ltd, and
printed and bound in Great Britain by
CPI Group (UK) Ltd, Croydon CR0 4YY

A CIP record for this book is available from the British Library

ISBN 978 1 4744 2781 4 (hardback)
ISBN 978 1 4744 2783 8 (webready PDF)
ISBN 978 1 4744 2784 5 (epub)

Contents

Acknowledgements

The idea of this book came about during the preparation of two panels on 'Shakespeare and science' for the Shakespeare 450 Congress held in Paris and organised by the Société française Shakespeare in April 2014. We are grateful to the Society, then presided by Dominique Goy-Blanquet, for its support.

We would also like to thank Collette Alexander for her help, as well as the two anonymous readers of Edinburgh University Press whose stimulating observations contributed to improving the first draft of this volume.

Lastly, we must single out Carla Mazzio, whose generous commitment allowed us to keep confidence in the whole project.

Textual Note

Unless otherwise specified, references to Shakespeare are to *The Complete Works*, ed. Stanley Wells, Gary Taylor, John Jowett and William Montgomery, Oxford: Clarendon Press, 2005 (2nd edn).

Notes on Contributors

Liliane Campos is a Senior Lecturer in English and theatre studies at Paris 3-Sorbonne Nouvelle, and co-founder of the university's 'science and literature' research seminar. Her current research interests focus on life sciences in contemporary drama and fiction. She has published articles about recent trends in European drama and British fiction and two book-length studies of science in theatre: *The Dialogue of Art and Science in Tom Stoppard's Arcadia* (Presses Universitaires de France, 2011) and *Sciences en scène dans le théâtre britannique contemporain* (Presses Universitaires de Rennes, 2012). She has also edited a special issue of *Alternatives théâtrales*, 'Côté sciences' (2009), and co-edited an issue of *Sillages critiques*, 'Le Théâtre et son autre' (2014), exploring the poetics of otherness in twentieth-century anglophone drama.

Sophie Chiari is Professor of early modern English literature at Clermont Auvergne University. She has written several books and articles on Shakespeare and his contemporaries. Her publications include *L'image du labyrinthe à la Renaissance* (Champion, 2010), *Shakespeare's Anatomy of Wit: Love's Labour's Lost* (Presses Universitaires de France, 2014) and *As You Like It: Shakespeare's Comedy of Liberty* (Presses Universitaires de France, 2016). She also edited *The Circulation of Knowledge in Early Modern English Literature* (Ashgate, 2015). She is currently working on court performances in Shakespeare's era.

Anne-Valérie Dulac is a Senior Lecturer in early modern literature at Paris 13 University. Her research interests include Sir Philip Sidney's works and correspondence, visual culture, limning, watercolours and the description and/or use of visual arts and artefacts in travel narratives and diplomacy.

Pierre Iselin is Professor of Elizabethan literature at the University of Paris-Sorbonne. He has published several monographs on Shakespeare's plays and numerous articles on stage music, language and the intellectual context of the debate on music in the sixteenth and seventeenth centuries. He is the musical director of an early music ensemble, 'The Sorbonne Scholars', in which he is also a performer.

Margaret Jones-Davies, former Senior Lecturer at Lille University (1970–85), then at the University of Paris-Sorbonne until 2012, has published several articles related to the philosophical context of Shakespeare's plays. She has worked on the influence of nominalism and of hermetic philosophy on Shakespeare's plays and prefaced eleven translations of Shakespeare's plays by André Markowicz (Les Solitaires Intempestifs). She has also contributed introductions to plays in the new Pléiade bilingual edition of Shakespeare (Gallimard, 2016).

Pierre Kapitaniak is Professor of early modern British civilisation at the University Paul-Valéry in Montpellier, France. His research bears on Elizabethan drama as well as on the conception, perception and representation of supernatural phenomena from the sixteenth to the eighteenth century. He is the author of *Spectres, ombres et fantômes: discours et représentations dramatiques en Angleterre* (Champion, 2008). He co-edited *Fictions du diable: démonologie et littérature* (Droz, 2007) and he edited and translated into French Thomas Middleton's play *The Witch/La sorcière* (Classiques Garnier, 2012). He is also engaged with Jean Migrenne in a long-term project of translating early modern demonological treatises into French, and has already published James VI's *Démonologie* (Jérôme Millon, 2010) and Reginald Scot's *La sorcellerie démystifiée* (Millon, 2015). He is currently working on the trilogy of Daniel Defoe's demonological treatises.

François Laroque is Emeritus Professor of English literature and early modern drama at Paris 3-Sorbonne Nouvelle. He has published several books on Shakespeare and Elizabethan drama, among which feature *Shakespeare's Festive World* (Cambridge University Press, 1991) and *Court, Crowd and Playhouse* (Thames & Hudson, 1993). He has also co-edited a two-volume anthology of *Elizabethan Theatre* (Gallimard, Pléiade, 2009) and published translations of Marlowe's *Doctor Faustus* and *The Jew of Malta*, and of Shakespeare's *Romeo and Juliet, The Merchant of Venice* and *The Tempest*. His last book, *Dictionnaire amoureux de Shakespeare* ('In love with Shakespeare. A personal dictionary'), appeared in February 2016.

Sélima Lejri is Lecturer in English and American literature and literary history at the University of Humanities and Social Science in

Tunis. She graduated from the University of Tunis and obtained a Masters degree (1998) and a PhD (2005) in Renaissance drama at Paris 3-Sorbonne Nouvelle. Her research interests include Greek, Renaissance, modern and postmodern drama, New Historicism and anthropology.

Frank Lestringant is Professor of French literature at the University of Paris-Sorbonne. Focusing on New World travel narratives and, more broadly, on sixteenth-century geography, he is the author of a great number of books and articles on the subject, including *Mapping the Renaissance World: The Geographical Imagination in the Age of Discovery* (Polity, 1994), *Cannibals: The Discovery and Representation of the Cannibal from Columbus to Jules Verne* (University of California Press, 1997), *Le Huguenot et le sauvage* (Droz, 2004), *Une sainte horreur, ou le voyage en Eucharistie (XVIe–XVIIIe siècles)* (Droz, 2013) and *Jean de Léry ou l'invention du sauvage* (Garnier, 2016).

Carla Mazzio teaches in the Department of English at the University at Buffalo, SUNY, where she is the Director of Graduate Admissions and Co-Director (with James Bono) of the Humanities Institute Science Studies Research Workshop. Her research focuses on literature and science, book history, and language development. She is the author of *The Inarticulate Renaissance: Language Trouble in an Age of Eloquence* (awarded the 2010 Roland Bainton Book Prize), the co-author of *Book Use, Book Theory: 1500–1700* (2005), the editor of *Shakespeare and Science* (2009), and the co-editor of three other books, including *Historicism, Psychoanalysis and Early Modern Culture* (2000) and *The Body in Parts: Fantasies of Corporeality in Early Modern Europe* (awarded the 1999 Beatrice White Book Prize).

Jonathan Pollock is Professor of English and related literature at the University of Perpignan-Via Domitia, France. He is the author of *Qu'est-ce que l'humour?* (Klincksieck, 2001), *Le Moine (de Lewis) d'Antonin Artaud* (Gallimard, 2002), *Le Rire du Mômo: Antonin Artaud et la littérature anglo-américaine* (Kimé, 2002), *Déclinaisons. Le naturalisme poétique de Lucrèce à Lacan* (Hermann, 2010) and *Lire Les Cantos d'Ezra Pound* (Hermann, 2014).

Mickaël Popelard is a Senior Lecturer in English studies at the University of Caen-Basse Normandie, France. He has written several articles on Elizabethan and Jacobean drama and early modern men of science such as John Dee, Thomas Harriot and Francis Bacon. He has published a book on Francis Bacon (*Francis Bacon: l'humaniste, le magicien, l'ingénieur*; Presses Universitaires de France, 2010) and a monograph on the figure of the scientist in Shakespeare's *The Tempest* and Marlowe's *Doctor Faustus* (*Rêves de puissance et ruine de l'âme: la figure du savant chez Shakespeare et Marlowe*; Presses Universitaires de France, 2010). He is currently working on voyages of exploration and discovery in the early modern period, and more particularly on Thomas Harriot's and John Davis's travel narratives.

Introduction

Sophie Chiari and Mickaël Popelard

So capacious and universal is Shakespeare's genius that it is often deemed capable of embracing every possible subject under the sun. And yet, by bringing together such different notions as science, technology, superstition and the theatre (i.e. the spectacular), the title of this book may very well strike the reader as something of an oxymoron. So we should perhaps start by saying that this volume makes the claim that, in Shakespeare's time, (what we now see as) science and (what we now call) literature were part of a common intellectual culture, which this volume aims at better mapping out. In order to do so, the following chapters will focus on a variety of Shakespeare's plays and scientific subjects with a view to bringing out the numerous and sometimes complex ways in which 'science', 'technology' and 'literature' may be said to have interacted in the early modern period.

But first, what is this thing called 'science'? As simple and straightforward as the question might sound, defining the term in such a way as to satisfy all – or at least most – philosophical schools, while still achieving a satisfactory degree of precision, may well prove an impossible task. If, for example, we adopt the popular inductionist view that science essentially consists of explanatory laws and theories which are based on, and ultimately derived from, factual observation, we run directly counter to Karl R. Popper's more elaborate *falsificationism* which posits that scientific theories are intellectual constructs which must be put to the test of experience. In Popper's epistemology, a good theory consists of refutable conjectures that, in all likelihood, will one day be disproved by experience. Conjectures and refutations (Popper 2000: passim), not observation, are what science ultimately boils down to. These, of course, are only two of the many possible definitions of science. It goes without saying that Popper's theory does not meet with unanimous consensus among epistemologists. One could also mention Thomas Samuel Kuhn, Imre Lakatos and Paul Feyerabend, as well as many other prominent thinkers who have put forward alternative, and sometimes conflicting, views of the nature and status of science.

Challenging the Literary/Scientific Divide

The paradox lies in the fact that, although we do have some difficulty in defining the term, few of us would probably deny that science has become a permanent and obvious feature of the world we live in. As Alan Chalmers explains in his short and classic study of the subject, we live in an age that tends to regard science as a modern religion.[1] So esteemed is science today that in recent years the term has often been used to add intellectual lustre and academic prestige to a whole range of subjects which have very little to do with actual science – at least if one chooses to adopt Popper's, Kuhn's, or indeed any other philosopher's definition of the term here expressed by Chalmers: 'Many areas of study are now described as sciences by their supporters, presumably in an effort to imply that the methods used are as firmly based and as potentially fruitful as in a traditional science such as physics or biology. Political sciences and social sciences are by now commonplace [. . .] In addition, Library Science, Administrative Science, Speech Science, Forest Science, Dairy Science, Meat and Animal Science and Mortuary Science have all made their appearance on university syllabuses' (Chalmers 1982: xix–xx). It looks as if one had to choose between defining science in either too specific or too vague a way – neither of which, of course, is satisfactory. According to John Worral in *The Routledge Encyclopedia of Philosophy*, 'what distinguishes science is the careful and systematic way in which its claims are based on evidence, [but] [t]hese simple claims, which I suppose would win fairly universal agreement, hide any number of complex issues' (Worral 1998: 573). There is every reason to believe, therefore, that any attempt at offering a normative and a priori definition of science should be abandoned in favour of a more empirical and naturalised approach, based on the actual practices of those 'knowledge practitioners' whom we would unanimously recognise as being 'scientists' in the true sense of the word.

---

[1] Chalmers 1982: xxi–xxii: 'A high regard for science is seen as a modern religion playing a similar role to that played by Christianity in Europe in earlier eras.'

In other words, while we may have difficulty defining science in a consensual, non-controversial way,[2] we are likely to agree about what truly qualifies as 'science' (genomics, or molecular biology, for example) and what does not ('Dairy Science', or literature and poetry, for that matter). Back in the late 1950s, C. P. Snow famously differentiated between two cultures, defining science negatively by separating it from what it was not, namely literature. In his younger days, he explained, scientists and what he called 'literary intellectuals' used to gaze at each other from across the two sides of a cultural ocean. But in postwar Britain, he continued, the gulf was even greater than it had been in his youth: 'thirty years ago, the two cultures had long ceased to speak to each other: but at least they managed a kind of frozen smile across the gulf. Now the politeness has gone, and they just make faces' (Snow 1993: 19). In his view, there were two types of intellectuals: the scientists and the literary intellectuals. Though he deplored such polarisation, he firmly believed that 'the intellectual life of the whole of western society [was] increasingly being split into two polar groups' (Snow 1993: 4). Between the members of the two groups, the communication was just as difficult – if not impossible – as if the scientists had been speaking only Tibetan and the literary intellectuals only English (or vice versa). By insisting there existed such a divide between two incompatible 'cultures', Snow offered a definition of science which coincided with some common-sense notions. According to him, science referred to the study of the physical world while literature concerned itself with more psychological and moral truths presented in a fictional form.[3]

It is hardly necessary to point out that Snow's theory creates more problems than it solves. For one thing, his 'two cultures' divide oversimplifies reality. As he himself acknowledged, the picture he painted is far too much of a sketch to be used as an accurate 'cultural map'.[4]

---

[2] See also Lewens 2015: 40–1. Failing to reach a definitive conclusion as to what distinguishes science from non-science, Lewens similarly argues that 'in spite of everything that we read about the importance of the "scientific method", it remains unclear what this method is [. . .] When we try to pinpoint some recipe for inquiry that all successful sciences have in common, we run into trouble.'

[3] Though Snow does not explicitly say so, such an opposition may certainly be inferred from *The Two Cultures*. For Snow's suggestion that literary intellectuals are not interested in the rational study of the physical world while scientists think that literature does not have anything to say about anything, see Snow 1993: 13–18.

[4] See Snow 1993: 10: 'I have thought a long time about going in for further refinements but in the end I have decided against. I was searching for something a little more than a dashing metaphor and a good deal less than a cultural map: and for those purposes the two cultures is about right, and subtilising any more would bring more disadvantages than it is worth.'

In fact, the 'two cultures' theory is not so much a fully fledged philosophical construction as it is a blueprint for educational and political reform. Besides, regardless of whatever validity Snow's theory might have had back in 1959, his views now sound very dated, influenced as they were by the context of the Cold War and the intellectual, social and political situation of the day. In any case, his description of a polarised society in which two groups of intellectuals stand staring blankly at each other in a state of mutual incomprehension sounds slightly passé today. The recent development of science and literature studies over the past thirty years or so is evidence enough that Snow's vision of a polarised society needs to be seriously reassessed. In 1959, Snow argued, not one in ten of the literary intellectuals he was acquainted with would have been able to define such basic scientific concepts as mass or acceleration.[5] Today, while some 'literary intellectuals' may still be rather uncomfortable with those concepts, one only has to flip through the books of any of the thinkers who have come to be (erroneously) amalgamated under the alleged banner of 'French theory' to realise that Snow's idea of a literary/scientific divide could hardly be further off the mark – although it is true that French theorists, from Lacan to Deleuze and Derrida, together with their followers, have sometimes been attacked, or even lampooned (most famously by Alan Sokal in the hoax article he submitted to *Social Text* in 1996),[6] for not fully mastering the scientific concepts they appropriated. And yet, as Alain Badiou's recent book suggests,[7] contemporary literary intellectuals and philosophers continue to hold science in general, and mathematics in particular, in very high esteem.

---

[5] See Snow 1993: 16: 'I now believe that if I had asked an even simpler question – such as, What do you mean by mass, or acceleration, which is the scientific equivalent of saying, Can you read?, not more than one in ten of the highly educated would have felt that I was speaking the same language.'

[6] See Sokal 1996: 217–52. For a discussion of Sokal's hoax and its intellectual repercussions in both France and the United States, see Cusset 2003: 13–23, about what Cusset calls the 'Sokal effect'.

[7] See Badiou 2015: 124. While Badiou believes that philosophy and mathematics still tend to be practised independently of one another, he calls for a tighter union to be formed between the two disciplines: 'in addition to the history of mathematics, one must arm oneself with philosophy. For it is in the mathematicians' interests to ask themselves what mathematics is about. And this is a truly philosophical question which can only be addressed philosophically' (my translation).

Epistemological Reflections on Early Modern Science

If there ever was a gap between science and literature, as people like Snow and Sokal contended in their respective ways, it is no longer the case, and scholars of the early modern period may certainly be said to have been instrumental in bridging it. In recent years, science has become one of the primary foci of a whole cohort of prominent historians who have turned their attention to various aspects of the early modern scientific world, from the study of scientific subjects at Oxford and Cambridge, to London instrument-makers and scientific communities, to women's supposedly privileged connection with the occult, not to mention monographs on such major intellectual figures as Francis Bacon, John Dee and Thomas Harriot. Thus, while Deborah E. Harkness has documented how London functioned as both 'a house of science and a prototype of a modern laboratory'[8] out of which a new approach to the natural world may be said to have emerged, Mordechai Feingold has established the positive contribution of early modern English universities to the shaping of modern science, challenging the received wisdom that they were 'repositories of traditional learning [. . .] inimical to scientific innovations' (Feingold 1984: 1). In his view, neither Cambridge nor Oxford can be adequately described as a conservative intellectual centre where every dangerously innovative idea was systematically and conscientiously nipped in the bud – or, to take a different and perhaps more appropriate metaphor, placed in a 'scholastic straitjacket' (Feingold 1984: 1–2). Mention, of course, should also be made of the groundbreaking work of such scholars and historians as Lorraine Daston, Katherine Park, Steven Shapin, Simon Schaffer, Paula Findlen, Peter Dear, Paolo Rossi, Brian Vickers, Robert Fox and Stephen Clucas, who, in their various ways, all helped to draw a much more accurate map of the early modern scientific world.[9]

---

[8] See Harkness 2007: 8: 'London, as a house of science and a prototype of a modern laboratory, worked. Despite the absence of a single institution to order and control it, her urban sensibility helped London practitioners successfully investigate nature, mediate conflicts over knowledge claims, collaborate on projects, expertly adjudicate disputes over methods, train new practitioners, seek financial support from civic and court figures, and negotiate their way through the challenges of studying nature in a crowded urban setting.'

[9] Some of the most important recent studies on early modern science in general and the 'scientific revolution' in particular include Daston and Park 1998; Pumfrey, Rossi and Slawinski 1991; Rossi 2001; Shapin 1994 and 1996; Shapin and Schaffer 1985; Dear 1995 and 2001; Smith and Findlen 2002; Vickers 1984; Clucas 2011. For monographs on the major figures of early modern science, with a special focus on England, see Fox 2000, Clucas 2006 and Sherman 1995.

If historians have recently become more and more interested in early modern science, as part of a more general shift towards intellectual and cultural history, so too have literary critics and Shakespearean scholars. Since the late 1960s and early 1970s, Shakespeare studies may be said to have branched out in almost every possible intellectual direction, influenced and revitalised as they were by such diverse theoretical currents and schools of thought as cultural materialism, feminism, deconstruction, new historicism and post-colonialism. At the same time, as students of Shakespeare and Renaissance drama were opening up new theoretical avenues, they also explored fields of knowledge that had so far received comparatively little critical attention. Science was one of these newly cultivated fields, with the result that we can now read erudite and stimulating book-length studies about how Shakespeare's drama reflects, dialectically relates to, or dynamically interacts with early modern medicine (Hoeniger 1992, Kerwin 2005, Pettigrew 2007), anatomy (Hillman 2007, Hillman and Mazzio 1997), geometry (Turner 2006) and the occult (Floyd-Wilson 2013), as well as seemingly more recent scientific concerns like cognitive sciences (Kinney 2004) and ecology and ecocriticism (Egan 2006). One should also mention scholars like Denise Albanese and Bernard J. Sokol, whose encompassing, or at least more broadly focused, studies paved the way for further research by stimulating scholarly interest in the interactions between Renaissance literature and early modern epistemology.[10]

To be fair, such a flurry of scholarly speculation about Shakespeare and science was not entirely unprecedented. As Carla Mazzio has shown in her illuminating introduction to a special double issue of the *Johns Hopkins Journal* devoted to Shakespeare and science, the late nineteenth and early twentieth centuries also witnessed a similar

---

[10] See in particular Sokol 2003 and Albanese 1996. As Sokol himself explains: 'some of my specific instances will point towards broad themes pertinent to Shakespeare's world: these will include the theme of new knowledge which overturns conventional wisdom; new knowledge which is hard to absorb or acknowledge; new knowledge based on approximations or otherwise known to be partial; and new knowledge based on the recognition and refutation of paradoxes. My purposes do not include philosophical discussion of such topics concerning knowledge, but rather an assessment of their impact on Shakespeare and his environment' (2003: 18–19). That Sokol was still unsure whether his attempt 'to pursue some little-studied epistemological themes of the play' which, so far, had only been 'tangentially' confronted (23) would be favourably received can be sensed, we think, both in his somewhat tentative tone and in his appeal for academic open-mindedness: 'disagreements need not be conducted in terms of caricaturing reductions or fervid denunciations of heresy' (19).

rise in books and studies approaching the Shakespearean canon from a variety of scientific and disciplinary perspectives, from meteorology and cosmology to anatomy and optics (Mazzio 2009: 5–6). Yet these studies often considered early modern scientific texts and ideas as some kind of intellectual substratum informing the literary content of the plays – together with many other similar influences, of course. In the words of Howard Marchitello, 'scientific texts were not "texts", as that word has come to be used in recent critical discourse, but sources. The model was background and foreground, and for students of literature the important works of science were "background"' (Marchitello 2006, quoted in Mazzio 2009: 18).

In contradistinction to such 'one-way traffic', this volume aims at offering a more dialectic vision of the Shakespeare/science nexus, taking up Mazzio's seminal idea that it is now necessary to 'move beyond forms of analysis focused largely on thematic traces, or indeed linguistic reflections, of historically specific arenas of scientific practice' (Mazzio 2009: 6). What we should pay attention to, instead, are the analogous 'procedures of thought', be they 'inferential, analytical, operative or hypothetical', which bind together scientific and literary discourses, practices and mentalities within a single episteme (in the Foucaldian sense of the word). Therefore, unlike late nineteenth- to early twentieth-century 'Shakespeare and Science' studies, this volume adopts a more complex and dynamic approach to the subject. But, distinct from the vast majority of recent attempts at shedding light on the interconnections between Renaissance drama and early modern scientific practices and discourses – which tended to privilege one particular type of scientific practice and/or discourse, be it magic, ecology or geometry – this collection deals with a wide range of scientific pursuits with a view to demonstrating the fundamental 'unity' of this 'province of knowledge' (in Bacon's words)[11] which Shakespeare inhabited and which constituted his creative and imaginative horizon.

In the introduction to his influential book *The Machine in the Text* (2011), Howard Marchitello proposes to distinguish between two types of methodological approach to the study of early modern literature and science, which he calls the 'influence' and the 'discursive' models. According to him and other critics like Elizabeth Spiller, for instance, the first wave of science and literature review, or 'influence model', was characterised by its propensity to treat scientific texts as a kind of backdrop against which early modern literature had to be read and

---

[11] See Francis Bacon's famous letter to Lord Burghley (1591) in which he writes that '[he] ha[s] taken all knowledge to be his province' (in Bacon 2002: 20).

interpreted. In this view, literature acquired a purely reflective status: early modern literary works were construed as mere 'commentaries on new scientific discoveries or intellectually interesting examples of the cultural work that literature might produce in the face of changing scientific knowledge' (Spiller 2007: 2, quoted in Marchitello 2011: 14).[12] It is precisely this one-directional 'influence model', Marchitello argues, which must be rejected in favour of the more dynamic 'discursive model' which gained momentum in the 1980s and 1990s. For the proponents of this second wave of criticism, literature and science should be treated in an egalitarian way and interpreted as different yet interrelated 'instances of early modern knowledge production' (Spiller 2007: 2). To put it differently, in order for such a model to be fully operative and functional, it is necessary to relocate science within its literary and cultural terrain. But it is equally important to regard art in general, and literature in particular, as contributing to the construction of early modern knowledge. As a consequence, such an interpretative model calls for the erasure of the science/literature polarisation and for the recognition of the socially and culturally embedded nature of both science and literature in the early modern period. The old hierarchical notion that science came first and literature second must be replaced by a more egalitarian approach whereby literary and scientific texts are placed on the same level and accorded equal epistemological importance, for, in the words of Marchitello, early modern literature and science ultimately share the same goal, namely converting 'personal experience into something like a universally valid truth' (Marchitello 2011: 15). The verticality inherent in the 'influence model' should therefore be discarded and replaced by a much more horizontal, dynamic and two-dimensional understanding of the relationship between science and literature. Following on from the work of Spiller, Mazzio and Marchitello, this volume therefore aims at shedding additional light on the exact nature of this two-way relationship by exploring a wide range of early modern practices and discourses, from alchemy to optics and from geography to medicine.

'What's in a name?': *Science* in Shakespeare's Age

Before we turn to the chapters that make up this volume, we should start by taking a closer look at the way the word 'science' was used by Shakespeare and his contemporaries so as to steer clear of

---

[12] See also Spiller 2009: 24–41.

anachronisms. It is important to realise that, as Steven Shapin once observed, the 'man of science was not a natural feature of the early modern cultural and social landscape. One uses the term *faute de mieux*, aware of its impropriety in principle, yet confident that no mortal historical sins inhere in the term itself' (Shapin 1998: 180). The same, perhaps, applies to that other problematic notion, 'science', which we also use, *faute de mieux*, while bearing in mind the necessary reservations and caveats above. Only by becoming aware of the risk of carrying into the past our modern acceptations of such words as 'science', 'art' or 'literature' will it be possible for us to understand the complex and subtle ways in which these early modern practices and discourses overlapped and/or interrelated. Yet because, in modern culture, the literary has come to be conceived of as 'the exotic other of the scientific' (Albanese 1996: 1), there is always a risk of one treating 'science' and 'literature' as if they were two separate and stable entities, 'two proud kingdoms lying alongside in chaste self-sufficiency' (Collini 1993: liii). It is worth remembering that such notions as 'science' and 'literature' are historically situated theoretical constructs. As the *OED* usefully reminds us, the word 'scientist' did not enter the English language until the 1830s.[13] Neither the Middle Ages nor the Renaissance recognised that there existed such a division between 'science' and 'literature', or 'art' and 'science' for that matter. In the words of Carla Mazzio, the word '"science" to Shakespeare's ears and eyes would have encompassed fields of knowledge including but also exceeding what was called, in the medieval period, the quadrivium (arithmetic, geometry, astronomy and music) and the trivium (rhetoric, grammar and logic)', so that 'as a disciplinary rubric "science" was often used interchangeably with "art"' (Mazzio 2009: 2). The 'seven liberal *arts*', for instance, were also referred to as 'the seven liberal *sciences*'. Derived from the Latin *scientia*, or knowledge, 'science' was simply coterminous with 'knowledge', so that any given form of knowledge could qualify as science, be it mathematics, history or grammar. But Mazzio goes on to argue that it is as dangerous to overemphasise the differences between 'art' and 'science' as it is to underplay them by reifying continuities between art (or literature) and science (Mazzio 2009: 2). While this is no doubt true, we would like to put forward

---

[13] See the following quotation from the 1834 article regarding the absence of a proper term to describe 'the students of the knowledge of the material world'. The author proposed that 'by analogy with artist, they might form "scientist"', though, according to the same author, 'this was not generally palatable' (see *OED*, s.v., 'science' meaning 5, quoted in Collini 1993: xii).

the view that continuities took precedence over differences and that, if terminology tells only part of the story, as Mazzio rightly insists, such lexical interchangeability does nonetheless indicate that early modern writers did not feel the need to differentiate between art, literature and science, or to treat them as separate categories.

As has already been pointed out, one of the central claims of this volume is that early modern writers did not distinguish between a 'scientific' and a 'literary' approach to the world they lived in and wrote about. This, of course, does not mean that they equated geometry with drama or poetry with alchemy. But they treated history, mathematics, philosophy and literature as different, yet equally respectable and interrelated, forms of knowledge. Poetry, for Sidney, was no less a 'science' than natural philosophy. Of course, some of Sidney's views were of a highly idiosyncratic character. Yet most of his contemporaries concurred that science and literature often went hand in hand: in short, there was as much poetry in knowledge as there was knowledge in poetry – so much so, in fact, that (what we identify as) 'art' and (what we would now see as) 'science' often seem to refer to overlapping, if not identical, practices. Prospero's magic is a case in point as it may serve to exemplify this blurring of the line between early modern 'art' and 'science'. Modern critics have tended to choose between two seemingly incompatible options, either identifying Prospero's 'rough magic' with theatre and illusion, or interpreting 'his potent art' in more scientific and magical terms. What we would like to suggest is that there need not necessarily be such a clear-cut opposition between these two readings of the play: if, scientist-like, Prospero manages to effect changes in the natural world by putting his theoretical knowledge to good practical use, his magic also takes the more artistic form of 'noises, / Sounds and sweet airs that give delight and hurt not'. Similarly, in the following extract from Bacon's *The Advancement of Learning*, 'art' is clearly construed as being synonymous with 'science'. The two words seem virtually interchangeable, since logic and rhetoric are successively referred to as 'the gravest of sciences' and 'the arts of arts': 'scholars in universities come too soon and too unripe to logic and rhetoric, *arts* fitter for graduates than children and novices. For these two, rightly taken, are *the gravest of sciences*, being *the arts of arts*; the one for judgment, the other for ornament' (Bacon 2000a: 59, our emphasis). Here, the art of rhetorical discourse and the science of logical discovery are clearly put by Bacon on a conceptual par. There is every reason to believe, therefore, that the either/or dichotomy should give way to a more inclusive approach whereby 'art' and 'science' become the two sides of the same early modern coin.

But technology must also be brought into the picture, for, as Carla Mazzio rightly points out, 'the relation among various practices of art and science were all the more marked, for "science," c. 1600, would have been aligned with spheres of labor, skill, or artisanal craftsmanship as diverse as writing (the craft of the scrivener), baking, brewing, husbandry, falconry, shoe-making, sewing, and surgery' (Mazzio 2009: 2–3). Therefore, in addition to referring to such disciplines as music, poetry or painting, 'art' could also mean 'craft', as noted by Carla Mazzio above, or 'science', as in the previous extract from Bacon's *The Advancement of Learning*, for example. But Bacon also equates 'art' with 'craftsmanship', emphasising the practical or technical connotation of the word, as opposed to its more theoretical meaning. This is especially evident in the following passage, where he argues that natural history would greatly benefit from the compiling of a history of trades and 'mechanical arts':

> But if my judgment be of any weight, the use of History Mechanical is of all others the most radical and fundamental towards natural philosophy; such natural philosophy as shall not vanish in the fume of subtle, sublime, or delectable speculation, but such as shall be operative to the endowment and benefit of man's life. For [. . .] it will give a more true and real illumination concerning causes and axioms than is hitherto attained. For like as a man's disposition is never well known till he be crossed, nor Proteus ever changed shapes till he was straitened and held fast; so the passages and variations of nature cannot appear so fully in the liberty of nature as in the trials and vexations of art. (Bacon 2000a: 65)

When theorising about science, Bacon weaves together strands belonging to several intellectual and/or practical traditions, including Renaissance humanism (in the form of references to classical rhetoric and literature) and artisanal craftsmanship. The alliance between 'the hand' and 'the mind' – i.e. practice and theory – lies at the heart of his reform of science, which aims at eschewing both idle speculation and blind empiricism. But to this definition of science one should also add a further dimension in the form of natural magic, for Bacon intends to incorporate a purified version of natural magic into his reform of science:

> Neither am I of opinion, in this history of marvels, that superstitious narrations of sorceries, witchcrafts, dreams, divinations, and the like, where there is an assurance and clear evidence of the fact, be altogether excluded. For it is not yet known in what cases and how far effects attributed to superstition do participate of natural causes; and, therefore, howsoever the practice of such things is to be condemned, yet from the

speculation and consideration of them light may be taken, not only for the discerning of the offences, but for the further disclosing of Nature. (Bacon 2000a: 63)

That natural magic should not be 'altogether excluded' from the realm of early modern science is made clear by Bacon's insistence that it, too, may shed light on the workings of nature. There is perhaps no better illustration of the three main intellectual streams that meet to constitute Baconian science than the portion of the *Gesta Grayorum* which is often believed to have been written by Bacon. In this text, which consists of speeches by six councillors about how best to rule a kingdom, Bacon has the second councillor advise the King that he should devote himself to 'the conquest of the works of Nature' (Marchitello 2011: 46).[14] In order to do so, the councillor argues that the King should set himself four main tasks: first, 'the collecting of a most perfect and general Library'; second, the designing of 'a spacious, wonderful Garden' with rare plants and animals, which will serve as 'a Model of Universal Nature made private' (ibid., 47); third, creating 'a goodly huge Cabinet, wherein whatsoever the Hand of Man, by exquisite Art or Engine, hath made rare in Stuff, Form, or Motion, whatsoever Singularity, Chance and the Shuffle of things hath produced, whatsoever Nature hath wrought in thing that want Life, and may be kept, shall be sorted and included' (ibid., 47–8); and fourth, building 'such a Still-house so furnished with Mills, Instruments, Furnaces and Vessels, as may be a Palace fit for a Philosopher's stone' (ibid., 48). The first and second tasks clearly point towards literary and scientific humanism, while the third is a reference to artisanal craftsmanship (as well as to Bacon's desire to compile a natural history of monsters). As for the fourth task, the phrase 'as may be a Palace fit for a Philosopher's stone' contains unmistakable echoes of natural magic and the practice of alchemy.

Shakespeare's science therefore lies at the crossroads between Renaissance humanism, technology and magic or, to use a word we moderns are more accustomed to, superstition. These strands were woven together within a common intellectual culture so as to compose a coherent whole. In that regard, *The Advancement of Learning* provides a good summary of the multifarious yet fundamentally unified and integrated nature of early modern science. It will perhaps

---

[14] For a detailed discussion of this text and the rest of the *Gesta Grayorum*, see Marchitello 2011: 24–50, 'Gray's Inn Revels, 1594–5'.

be argued that Bacon is the last author we should turn to if we are to determine what Shakespeare's contemporaries thought about knowledge or science. Bacon's work, one will object, is too exceptional and revolutionary for it to reflect the most commonly held beliefs of his time. Yet Bacon often seized on received ideas as a basis on which to build his philosophy, and although his plan for the reform of science is quite revolutionary, his terminology often remains rather traditional. Besides, as will become apparent in the course of this introduction, Bacon's philosophy is a common thread running through several chapters of this volume. The following passage deserves to be quoted in full because it provides a clear illustration of what a multifaceted notion 'science' was in the early modern period:

> But because the distributions and partitions of knowledge are not like several lines that meet in one angle, and so touch but in a point, but are like branches of a tree that meet in a stem, which hath a dimension and quantity of entireness and continuance before it come to discontinue and break itself into arms and boughs; therefore it is good, before we enter into the former distribution, to erect and constitute one universal science, by the name of 'Philosophia Prima,' Primitive or Summary Philosophy, as the main and common way, before we come where the ways part and divide themselves [. . .]
>
> For example; is not the rule, 'Si inqualibus æqualia addas, omnia erunt inæqualia,' an axiom as well of justice as of the mathematics? and is there not a true coincidence between commutative and distributive justice, and arithmetical and geometrical proportion? Is not that other rule, 'Quæ in eodem tertio conveniunt, et inter se conveniunt,' a rule taken from the mathematics, but so potent in logic as all syllogisms are built upon it? Is not the observation, 'Omnia mutantur, nil interit,' a contemplation in philosophy thus, That the *quantum* of nature is eternal? in natural theology thus, That it requireth the same omnipotency to make somewhat nothing, which at the first made nothing somewhat? according to the Scripture, 'Didici quod omnia opera, quae fecit Deus, perseverent in perpetuum; non possumus eis quicquam addere nec auferre.' Is not the ground, which Machiavel wisely and largely discourseth concerning governments, that the way to establish and preserve them is to reduce them *ad principia*, a rule in religion and nature, as well as in civil administration? Was not the Persian Magic a reduction or correspondence of the principles and architectures of nature to the rules and policy of governments? Is not the precept of a musician, to fall from a discord or harsh accord upon a concord or sweet accord, alike true in affection? Is not the trope of music, to avoid or slide from the close or cadence, common with the

trope of rhetoric of deceiving expectation? Is not the delight of the quavering upon a stop in music the same with the playing of light upon the water? (Bacon 2000a: 76–8)

In the 'stem' of his tree of knowledge, Bacon gathers together axioms borrowed from the law, mathematics, logic, natural philosophy, theology, biblical studies, politics, magic, music and rhetoric. There are 'rules' and 'principles' which are common to all intellectual disciplines, be they 'literary' or 'scientific', to use anachronistic labels. Laying bare some of these 'principles' while studying the ways in which they may be said to inform and interact with Shakespeare's drama is precisely what this volume aims at. 'Spectacular science', the title we chose for this book, enhances a series of complex and dynamic relationships between two 'proud kingdoms', to borrow Collini's phrase (1993: liii), namely that of Shakespeare's drama and that of early modern science, the latter being construed as an intellectual territory in which art, technology and superstition overlap to such an extent that they become virtually indistinguishable. Because early modern science and literature were entwined and embedded within a common culture, like two branches shooting out from a common intellectual stem, the following chapters will analyse some of the 'rules', 'maxims' or – in Carla Mazzio's words – 'thought-processes' underlying both early modern science and Shakespeare's literary art, while insisting that the relationship between Shakespeare and science must be understood as being 'discursive' and dynamic, rather than static and unidirectional.

Shakespeare's Spectacular Uses of Science and Technology

By the second half of the sixteenth century, more and more poets and playwrights started exploring the tricky relationships between sensory experience and scientific explanation. Shakespeare's own interest in science, as well as his literary and dramatic use of natural phenomena, has been clearly established. Comparable to Roger Bacon and his talking brass head in Greene's *Friar Bacon and Friar Bungay*, or to Marlowe's Doctor Faustus, Shakespeare's magus, Prospero, is 'for the liberal arts / without a parallel' (1.2.73) and yet he is also a flawed human being whose magic art seems linked to his artistic imagination. To some extent, he can be compared to John Dee, who determined the coronation day of Queen Elizabeth and who was renowned for his magic *scrying*[15] mirror, a polished disc of black obsidian imported from Mexico, which testified to the Elizabethan fascination with cosmology and astrology.

This fascination seems omnipresent in early modern drama. This should come as no surprise since early modern science and technology incorporated a number of performative, or spectacular, elements which rendered them attractive not only for the literati but also for the lesser educated.[16] A striking example of this is Dee's creation, in 1547, of a flying mechanical beetle for a school drama performance at Trinity College, Cambridge (Szönyi 2004: 114). Dee was very proud of his spectacular device and, as he reported the event in his *Autobiographical Tracts*, he significantly focused as much on the technical feat he had accomplished by resorting to Euclidean geometry as on the sense of wonder he had caused among the audience:

> Hereupon I did sett forth (and it was seene of the University) a Greeke comedy of Aristophanes, named in Greek *Eirene*, in Latin *Pax*; with the performance of the *Scarabeus*, his flying up to Jupiter's palace, with a man and his basket of victuals on her back: whereat was a great wondering, and many vaine reports spread abroad of the means how that was effected. (Quoted in Reilly 2011: 26)

Stagecraft and science appear strongly interrelated here, for Dee's mechanical marvel was clearly meant to be displayed. In other words, knowledge and theatrical practices naturally informed one another in the early modern era. It would be fair to say, like Adam Max Cohen, that '[t]he theatre seems a natural place to seek out representations of technologies because of the cross-fertilization between technological and theatrical imagery' which was clearly at work in the numerous scientific treatises issued in the course of the sixteenth century (2012: 705). In fact, the increasing 'mathematicization of nature' (Crombie 1990: 158) at the turn of the seventeenth century went hand in hand with the mathematicisation of the stage. Shakespeare's plays, in particular, include a number of special effects, and his more mature works, especially conceived for the private theatre, capitalise on 'the acoustic and

---

[15] 'Scrying' may be defined as the practice of looking into a translucent ball or disc with the belief that angels may be seen in it. On this, see Harkness 1999.

[16] On the spectacular dimension of early modern science, see Dawbarn and Pumfrey 2004. Though they argue that 'ostentatious' or spectacular science was more commonly encountered in Continental Europe, it was also a feature of English scientific displays, as the examples of John Dee or Cornelis Drebbel clearly suggest.

optical properties' of indoor playhouses 'to test the possibilities and limits of the greater control over audience perspective afforded by newly popular forms (like the masque) and venues (like Blackfriars)' (Crane 2013: 264). It would be wrong, though, to think that science and technology only appear in such late plays as *Cymbeline*, *The Winter's Tale* and *The Tempest*, for it is Shakespearean drama in general which draws upon less tangible, if no less real, scientific concerns in order to move, amaze and, perhaps, slightly disturb playgoers. It pays special attention, for example, to the ever-growing divide between sight and the external world and it raises nagging questions related to the (mis)use of science. More often than not, tragedies, histories and comedies alike seem permeated with the disturbing themes of epistemological uncertainty and nothingness. Should we see the new emerging science as an agent of positive transformation? Or should we rather stick to superstition as a cohesive strength uniting society? To the first question, the playwright responds by staging deeply rooted anxieties about the capacity of mankind to reinvent itself and to adapt to new horizons without damaging its environment and betraying its humanist ideals. To the second one, he clearly answers that beliefs, based on fear, credulity and distrust, tend to disconnect people more than bring them together.

Yet, in Shakespeare's era, science and technology, which put forward the use of reason and method to the detriment of popular wisdom, cannot be totally detached from belief.[17] Weather predictions, for example, could supposedly be achieved through a thorough observation of nature and a keen perception of God's power; medicine implied a good knowledge of anatomy as well as of astrology; and, more generally, superstition could seldom be reduced to ignorance. As more often than not the occult and the scientific were still narrowly linked, Shakespeare, in his plays, walks this rather thin line in order to question both traditional paradigms (the theory of humours, the music of the spheres, astrological predictions, and so on) and the new forms of 'scientific', or 'rational', thinking.

It goes without saying that this one volume does not claim to deal with all of the scientific themes and techniques mentioned by Shakespeare in his plays and poems. Had we intended to do so, a wide range of subjects should have been covered, going from mathematics

---

[17] It might be argued, though, that even contemporary science does not preclude belief entirely, for both theories and the observation statements they are based on are of a conjectural nature, as Popper explained. Far from 'resting upon solid bedrock', science therefore 'rises above a swamp' (Popper 1992: 94). For a discussion of the 'theory-ladenness of observation' and science as 'piles in a swamp', see Lewens 2015: 30–4.

(Navarre's inhabitants, in *Love's Labour's Lost*, keep reckoning numbers but betray poor mathematical abilities) to climatology (Titania, in *A Midsummer Night's Dream*, famously itemises a whole catalogue of disasters, including fog, flooding, failed crops, miserable flocks, polluted air, bad temperatures and rheumatic diseases, all due to the wet climate; 2.1). Our current project is rather to depict Shakespeare's ambiguous attitude towards science and to underline the fairly broad view of early modern science and technology which he presents while simultaneously emphasising his difficulty in clearly distinguishing between popular beliefs and innovative knowledge. In fact, the playwright incorporates both aspects in his work and he leaves us free to decide. Contrary to Marlowe, Jonson or, say, Middleton, who offered rather univocal viewpoints on the early modern world, Shakespeare provides us with a variety of perspectives which propose, but never impose, possible (if questionable) systems of thought. What is sure, however, is that the underlying presence of several aspects of Bacon's philosophical interests in Shakespeare's texts shows how attentive to the 'scientific revolution' of his time the playwright was. This was a time when magic, religion and science were just beginning to divide and Shakespeare resorted to scientific (i.e. geographical, medicinal, optical, etc.) discourses to comment on the (dis)functioning of the body politic. In other words, 'science' was a way for him to enlarge our perception of the power relationships in his work as his figures of knowledge are often invoked at crucial moments in the plays where they are either feared or mocked.

Indeed, for all his charisma, the early modern 'scientist' remained a disquieting character. As noted by David Hawkes, 'Bacon's trope of the scientist as torturer', for instance, 'is invoked in various [Shakespearean] contexts and evaluated in a variety of manners' (Hawkes 2014: 24). The increasing power of early modern men of science posed the question of the limits of the human individual, and if it did propose solutions to improve the future, it also presented the individual with new dangers. However, such anxiety was not enough to destroy the general fascination exerted by the promoters of those new modes of thinking. 'There are more things in heaven and earth', Hamlet tells Horatio, 'than are dreamt of in our philosophy' (1.5.168–9). Besides his endorsement of the supernatural, Hamlet's remark may also indirectly testify to Shakespeare's growing obsession with mathematics and astronomy after new paths had been opened by the invention of the telescope. The playwright was indeed probably acquainted with Thomas Digges (c. 1546–95) – the father of Leonard Digges (1588–1635), who wrote one of the prefatory poems for the 1623 Folio – who may have kept him posted about the latest findings of Copernican philosophy. In *A Perfit Description of the Caelestial Orbes*, a free translation of Book I of *De revolutionibus orbium caelestium* (1543) published in Leonard Digges's 1576 *Prognostication euerlastinge*,[18] Thomas contributed to the spreading of Copernicus's theories in England,[19] and these were brought one step further by Galileo Galilei, who was the playwright's exact contemporary and who died twenty-six years after him. In 1610, Galileo published *Sidereus Nuncius*, a treatise in which he reported his discovery of Jupiter's satellites which represented a major breakthrough, allowing him to conclude definitely that our universe was heliocentric and not, as long believed before him, geocentric. This conclusion was adopted – not without a certain degree of scepticism – by a number of early modern thinkers and writers, who were intrigued by the translunary world and who wanted to widen their intellectual horizon. If Imogen's declaration in *Cymbeline* – 'O, learn'd indeed were that astronomer / That knew the stars as I his characters' (3.2.27–8) – primarily alludes to 'astronomers' in the sense of 'astrologers', it probably also refers to Galileo's observations, thus conflating traditional wisdom with new scientific discoveries.

The 'scientific revolution', as Shakespeare and his contemporaries probably experienced it, did not erase the knowledge of the past – far from it. It simply reappropriated it in a new way, putting to the fore the most innovative thinking of the Ancients. No wonder, then, if some of Shakespeare's characters seem to be aware of Epicurean philosophy. One thinks here of Hamlet, Lear or Duke Vincentio in *Measure for Measure*. Jonathan Pollock even goes as far as to suggest that Shakespeare had probably read Lucretius at first hand,[20] a bold assertion

---

[18] This Leonard Digges (c. 1515–c. 1559) was the grandfather of Leonard Digges and the father of the mathematician Thomas Digges. He wrote *A prognostication of right good effect*, apparently first printed in 1553, but whose earliest extant edition dates from 1555. This text was reprinted a number of times – at least thirteen until 1600 – and '[f]rom 1576, the work became the vehicle for an addition by his son Thomas, which included a presentation of Copernicus's heliocentric world system'. See Johnston 2004a.

[19] Thomas Digges notably included 'a famous diagram which went beyond Copernicus's own scheme, by showing an infinite universe in which the stars extended indefinitely outwards from the solar system'. See Johnston 2004b.

[20] Indeed, it seems now 'possible to class the reception of Lucretius in England up to the time of Shakespeare into three distinct categories, depending on whether the native writers allude to the poetry, ethics or physics of *De rerum natura*' (Pollock 2015: 48).

which would then explain why several references to atomism are found in plays written much earlier than, say, *King Lear*, where Lucretius is alluded to via Florio's translation of Montaigne's *Essays* in 1603. In *As You Like It*, for example, when Celia ironically tells Rosalind that '[i]t is as easy to count atomies as to resolve the propositions of a lover' (3.3.194–5), this may well be an indirect allusion to Lucretius's theory of atomism. Celia's remark 'turns the art of love into an uninterpretable object in a setting of profound meaninglessness, drawing ever-so-slight attention to the intersection of macrocosmic nature and the human psyche, the moral trajectory of atomism' (Lepage 2012: 87). More interestingly, while 'Epicurean morals', frequently associated with 'self-indulgent pleasure-seeking and atheism', clearly 'had a bad press' in Shakespeare's England (Pollock 2015: 48), Epicurean philosophy is here apparently only endorsed by Rosalind's cousin to comment sarcastically on the changeability of love's vows as well as to underline the fact that we live in a universe in constant motion.

Yet, as is usual with Shakespeare, the playwright's exact position on science, technology and superstition is difficult to assess. From play to play, his attitude varies and it is impossible to trace an evolution that would, say, go from a faith in traditional wisdom to a conversion to more 'scientific' forms of knowledge. In *Romeo and Juliet*, for example, he presents us with '[a] pair of star-crossed lovers' (Prologue, l. 6) while never questioning the validity of this assumption in the play, and in *King Lear* the bastard describes astrology as pure nonsense and deception.[21] Shakespeare's frequent use of optics is similarly disconcerting. Whereas the arch-villain Richard III intends to be 'at charges for a looking-glass' (1.2.259) in order to 'adorn' his deformed body (261), the poet of Sonnet 62 looks at himself in the mirror, only to underline the discrepancy between his supposedly perfect face and his real features ('But when my glass shows me myself indeed, / Beated and chopped with tanned antiquity, / Mine own self-love quite contrary I read'; ll. 9–11). The use of medicine is marked by a similar volte-face if we note that, in *All's Well that Ends Well*, we are led to believe that 'some prescriptions / of rare and proved effects' allow Helen to heal the 'desperate languishings' of a sovereign (1.3.221–30), while we are

---

[21] See Edmund's tirade in *King Lear*, 1.2.119–26: 'This is the excellent foppery of the world, that when we are sick in fortune, often the surfeits of our own behaviour, we make guilty of our disasters the sun, the moon and the stars, as if we were villains on necessity, fools by heavenly compulsion, knaves, thieves and treachers by spherical predominance; drunkards, liars and adulterers by an enforced obedience of planetary influence; and all that we are evil in by a divine thrusting on.' See Laroque's chapter in the present volume, p. 40.

forced to acknowledge that, in *King Lear*, Cordelia's proposal to cure her 'child-changed father' (4.7.17) thanks to the sounds of music is bound to fail. Such ambivalence illustrating Shakespeare's reluctance to choose between the old and the new episteme is sometimes found in one and the same play. Is the playwright serious, for example, when, in *As You Like It*, he has Touchstone describe Corin, a man whose wisdom is merely composed of proverbs, as a 'natural philosopher' (3.2.30)? Or does he deride traditional wisdom to promote instead a new order of things embodied by the sarcastic Fool in the play?

It was then up to the spectators to decide and provide their own answers, so that the staging of this dilemma was an efficient way to make the debate popular and give the audience food for thought. Brimming with innovation and tradition, new techniques and ancient learning, the stage was also an imaginative cell in which the actors strived to apply the more recent technical discoveries. Theatrical issues in connection to science, technology and superstition are therefore among the main concerns of the following chapters. For much as 'science' can be said to have shaped at least part of Shakespeare's dramatic writing, conversely, the Elizabethan stage was also a site of knowledge as well as entertainment, an arena where science suffered a 'sea-change' and was turned into an art. It is perhaps not exaggerated to consider the stage as a laboratory of sorts, in which scientific experiments were of course not actually produced but could at least transform ordinary perceptions into 'something rich and strange' (*The Tempest*, 1.2.404).

An Overview of the Volume

This volume is divided into four broad areas encompassing the interrelated concepts of science, technology and superstition in the Shakespearean canon, namely popular beliefs, healing and improving, knowledge and (re)discoveries, and mechanical tropes. Science in Shakespeare's time simultaneously occupied a material, public space and a metaphorical, private one, and through an understanding of these two dimensions, these chapters endeavour to bridge the gap between them. The main objective is made clear from the very first contribution: their purpose is to reassess the sometimes contradictory ways in which early modern scientific practices are reappropriated by Shakespeare in most of his works, including his poetry.

In the first section, devoted to popular beliefs, two chapters examine the tenuous line separating science from the beliefs and intellectual traditions inherited from the past. Astrology and demonology are here examples of cultural practices whose inconsistencies were already pointed at and sometimes exposed in early modern England, so that they were very gradually relegated to the margins of knowledge. François Laroque starts by exploring the paradoxical purposes of the 'science' of astrology in two plays, namely *Romeo and Juliet* and *The Tempest*, and then in the sonnets. While a few sceptics continued to deride astrology, there was a striking revival of this science by the end of the sixteenth century, as influential mathematicians like John Dee and Thomas Digges contributed to popularise it. What then became a real craze for astrology clearly found its way into Shakespeare's poetry. In Sonnet 14, for example, where the speaker deliberately poses as a mock-astrologer, and in Sonnet 107, he playfully redefines what 'poetic' astrology should be, as his own astrological skills enable him to read the eyes of the young man as if they were two fixed stars. In the plays, it can even be a structural device when, as Laroque argues, the 'ancient grudge' of the Montagues and the Capulets in *Romeo and Juliet* is marked in the initial prologue by its astrological connotations. The role of the stars is of prime importance in the tragedy and it makes us understand the tragic destiny of the title characters who are doomed to death and destruction. Interestingly, Shakespeare kept thinking about the influence of the planets throughout his career, for a much later play like *King Lear* is similarly concerned with stars and disasters. The vivid opposition between Gloucester and Edmund allows Laroque to demonstrate that the playwright was interested in the vivid controversies over the validity of astrology, even though Shakespeare refuses to take sides.

Pierre Kapitaniak follows up on this as he reassesses Shakespeare's understanding of the links between science and local superstition in an analysis of the role of witches and witchcraft in his plays. Despite legends about King James I ordering it to be burnt, Reginald Scot's *The Discoverie of Witchcraft* (1584) was an ongoing success from the time of its publication, more often meeting with approval than with condemnation. Several generations of London playwrights were indeed among those who approved of Scot's ideas. In *The Discoverie of Witchcraft*, they found elements of inspiration for their supernatural figures that became quite successful on the Elizabethan and Jacobean stages, and one can only wonder whether the slow evolution from the usual supernatural paraphernalia (ghosts, demons, witches and wizards) towards more and more unbelievable figures is not due to Scot's influence. Kapitaniak tries to reassess whether undisputable traces of Scot's treatise can be found and ascertained in Shakespeare's plays and, if so, what sort of hypotheses all this may offer for the understanding of the playwright's attitude towards science and superstition.

Part two, 'Healing and Improving', stems from the idea that early modern scientists often envisaged science as a means of manipulating nature and society to useful ends and, as such, as a tool likely to improve people's lives – even though, with the benefit of hindsight, one could easily notice that the destructiveness of scientific practices so often denounced nowadays already started to characterise the budding modern science. Sélima Lejri, in her chapter, sheds fresh light on the emerging scientific etiologies propounded by the physicians of the time. She shows that it is thanks to Edward Jorden's *A Briefe Discourse of a Disease Called the Suffocation of the Mother* (1603) that the interpretations of demonic vexation started to give way to the rational alternative of hysteria. Shakespeare's interest in the medical theories of physiology, mainly humoralism, then became palpable due to the influence of Timothy Bright's or Edward Jorden's ideas. Within this context, the witch-hunting period came to an end and gave credit to reason over superstition, and Shakespeare's representation of the female body in his Jacobean plays bears the contemporary stamp of these new sources of information. It is his response to such contemporary scientific theories that Lejri's chapter tackles via the example of *hysterica passio*, a female pathology much discussed at the time and explicitly mentioned in *King Lear*.

In the next chapter, Pierre Iselin broadens this clinical picture of the early modern society by calling attention to one of the great diseases of the time, i.e. melancholy, either cured or aggravated by music, depending on the seriousness of the patient's illness. Iselin then chooses to deal with early modern music in a play like *Twelfth Night*, one of Shakespeare's most musical comedies, within the polyphony of discourses – medical, political, poetic, religious and otherwise – on appetite, music and melancholy, which circulated in early modern England. Commenting on the supposed soothing power of music, Iselin examines how these discourses interact with what the play actually says about music in the many dramatic commentaries of the text, and what music itself says in terms of the play's poetics. The abundant music of *Twelfth Night* is considered not only as 'incidental' but as a sort of meta-commentary on the drama and the limits of comedy. Pinned against contemporary contexts, *Twelfth Night* is therefore regarded as an experiment in the aural dimension and as a play in which the genre and mode of the song, the identity and status of the addressee, and the more or less ironical distance that separates them constantly interfere. Ultimately, the author sees in this dark comedy framed by an initial and a final musical event a dramatic piece punctuated, orchestrated and eroticised by music, whose complex effects work both on the onstage and the offstage audiences.

Now, if music could help men to improve their mental condition, the same was true of alchemy, which, as Bacon pointed out, was poised between knowledge and superstition and which, above all, aimed at perfecting nature. In her chapter devoted to the 'abstract riddles' of alchemy (Ben Jonson, *The Alchemist*, 2.1.104), Margaret Jones-Davies argues that Shakespeare uses the poetics of alchemy at the moment when it begins to be on the wane as a 'science'. Indeed, at the turn of the sixteenth century, the literal reading of the power of alchemy was being questioned. And yet, no matter how cruel the satire and the persecutions against some alchemists had become, the influence of alchemy remained active on a figurative level, just as hermetic philosophy acquired a political importance as the basis of a new religious language freed from the fanaticism of the warring parties. Now, if Shakespeare somehow happened to share Rabelais's and Erasmus's irony against alchemical lore, he did not extend his scepticism to the ideal of perfection, which he expressed through alchemical imagery and numerology, more particularly in the histories and the romances. Alchemy doesn't work miracles, Jones-Davies notes, but by lending its language to the ideal of perfection, it certainly creates wonder.

The third part of the volume deals with the dramatic interplay between knowledge and scientific (re)discovery. In this part, 'science' becomes synonymous with 'knowledge or understanding acquired by study' (*OED* 2), while also retaining most of the connotations associated with the other senses of the word as analysed in parts one and two. For if Shakespeare was not a 'University Wit', he was definitely an educated reader able to make the most of the books he had the opportunity to read. Interestingly, in a number of plays, he alludes to the Epicurean atomist theory that is the subject of Jonathan Pollock's chapter. Even though the playwright had no access to the writings of Epicurus, it is likely that he knew the *De rerum natura* by Lucretius, were it only via Montaigne's *Essays*. It is Pollock's contention that the prevalence of weather imagery in Shakespeare's later plays is the result not just of his propensity for cloud-gazing but also of his interest for Lucretius's use of meteorological models that explain the creation and disintegration of material objects and of living beings. Epicurean science recognises only (atomic) matter and the void and denies the reality of a spiritual 'substance' (God or an immortal soul). It would seem that Shakespeare uses the Lucretian doctrine as a means of establishing dialectical oppositions: set against Lear's naive paganism or Cordelia's redemptive figure, for instance, atomism portrays a world deprived of divine providence and subject to purely material forces.

Also exploring Shakespeare's borrowings from scientific sources, Anne-Valérie Dulac turns to optics and takes *Love's Labour's Lost* as a case for study. The author first reminds us that in her *Study of* Love's Labour's Lost, published in 1936, Frances Yates repeatedly mentions the importance of Alhazen's optical theory through the play's many references to light, eyes and vision. Dulac then proceeds to deal with two mistakes made by Yates in her rather short description of the 1572 edition of the *Opticae thesaurus*, a compendium including a truncated Latin version of Alhazen's treatise along with Witelo's *Perspectiva*. She demonstrates that this was due to the fact that, at the time Yates was writing, historians of science had not yet shown how different the translations of the *Kitab al-Manazir* (*Book of Optics*) are, or, in other words, how different Alhazen is from Alhacen and Ibn al-Haytham. Finally, Dulac looks into the Latinised version of Alhazen's optical theory to inquire whether it could explain some of the most intricate metaphorical networks of Shakespeare's early comedy of wit.

If this study is devoted to a play which is much concerned with the power of words and language in general, Frank Lestringant turns to a late tragicomedy, *The Tempest*, whose rich background reveals a fascination with travels, explorations and overseas discoveries. Lestringant shows how the playwright incorporates science, understood here as a vast area of knowledge encompassing several branches of learning. Indeed, in *The Tempest*, books (grammar, logic and rhetoric forming the medieval trivium) and maps (geometry being part of the quadrivium) are essential elements that emphasise the role of education and knowledge in a changing world. Like Pollock, Lestringant reminds us of Shakespeare's debt to Montaigne's *Essays* in Florio's translation, and of the way he cleverly reappropriates Montaigne's 'negative formula' in his essay entitled 'Of Cannibals'. He focuses on the way Shakespeare manages to dramatise Montaigne's observations and on how he lionises old Gonzalo thanks to a number of indirect quotations. Doing so, he reassesses the old counsellor's role in *The Tempest* and he rehabilitates both his knowledge and humanist training.

Mickaël Popelard ends this third part with an analysis of the rising interest in the endless transformation of nature as exemplified by Bacon's philosophy and Shakespeare's *Tempest*. He takes a look at the making of early modern science and provides epistemological considerations on Shakespeare's approach to limits and the unlimited, thus taking up some of the issues dealt with in Pollock's and Jones-Davies's chapters. Taking *Macbeth*'s idea of an essentially limited human nature as his point of departure, Popelard first focuses on Bacon's active and practical stance, insisting on the fact that, for him, the role of the scientist consists in binding theory and practice together so as to achieve 'the effecting of all things possible'. While he posits that Bacon's reform of science and philosophy is marked by its open-endedness and, therefore, by its absence of limits, he also shows that, if a similar interest in boundlessness can be noted in Shakespeare's late plays, characters such as Prospero remain constrained by their obsession with 'limits', 'confines' or 'boundaries'. Yet, for all his epistemological hesitancy, Shakespeare's magician and/or natural philosopher shares some of Bacon's ideas on science, the most important being the belief in an operative rather than a verbally oriented science.

In the fourth and last section, entitled 'Mechanical Tropes', two chapters address the question of technological advances in Shakespeare's time and deal with particular objects. Sophie Chiari first probes the question of time technology in an analysis of clocks and dials as they are referred to in a number of plays. If the lower classes could only measure time thanks to public sundials, the better-off benefited from a direct, tactile contact with the new instruments worked out to allow a more precise reckoning of time. If owning a miniature watch was still a privilege at the end of the sixteenth century, Shakespeare does record this new and expensive habit in his plays. Dwelling on the anxiety of his wealthy Protestant contemporaries, thereby emphasising their fear of the *eschaton*, the playwright indeed pays attention to the materiality of the latest time-keeping devices of his age, sometimes introducing unexpected dimensions in the art of measuring time. Moreover, the mechanisation of time was also a means of reminding people of their last end, so that mechanical clocks were indirectly linked to the medieval trope of *contemptus mundi*. The new time technology, Chiari suggests, led to a disquieting fashioning of the early modern self, partly baffled by the discovery of new lands, new truths and new horizons, and yet quite obsessed with death and physical decay.

Liliane Campos's chapter on Stoppard's *Rosencrantz and Guildenstern Are Dead*, a play first staged in 1966, concludes this final part which bridges the gap between early modern and modern science, thereby opening up new interpretative possibilities. By decentring our reading of *Hamlet*, Campos argues that Stoppard's tragicomedy questions the legitimacy of centres and of stable frames of reference. She examines how Stoppard plays with the physical and cosmological models he finds in *Hamlet*, particularly those of the wheel and the compass, and thus gives a new scientific depth to the fear that time is 'out of joint'. In both his play and his own film adaptation (1990), Stoppard's rewriting gives a twentieth-century twist to these metaphors through references to relativity, indeterminacy and the role of the observer. When they refer to the uncontrollable wheels of their fate, his characters no longer describe the destruction of order, but the uncertainty as to which order is exactly at work, whether heliocentric or geocentric, random or tragic. When they express such loss of their bearings, they do so through the experiments of modern physics, from Galilean relativity to quantum uncertainty, drawing our attention to shifting frames of reference. Much like Schrödinger's cat, Stoppard's Rosencrantz and Guildenstern are both dead and alive. As we observe their predicament, Campos argues, we are placed in the paradoxical position of the observer in twentieth-century physics, and we are constantly reminded that our time-specific relation to the canon inevitably determines our interpretation.

The volume comes full circle with Carla Mazzio's coda, which pivots on the opening words of our title, 'Spectacular Science', in order to reconsider questions of vision and cognition and spectacularism and scepticism at stake in the volume more broadly. Since science and technology often relied upon theatrical or spectacular modes of making and representation, she observes, it is important to ask the further question: just what did the theatrical spectacle do to forms of science already fashioned *as* spectacle, on the one hand, and to forms of science that were distinctly unspectacular, otherwise elusive, that is, to embodied experience or direct sensory perception, on the other? Mazzio evokes multiple historical frameworks to consider how early modern dramas may have worked to render 'science' at once more accessible and more subject to sceptical inquiry. By focusing on the logic of sensation undergirding a number of chapters in this volume, moreover, Mazzio reassesses the contents of the book in a way that highlights the work of theatre in relationship to knowledge acquisition and the agency of the spectator, auditor and otherwise multi-sensory being in the early modern theatre. She closes the coda by considering politics of knowledge onstage as we come to understand how early modern audiences contributed, in complex and multi-layered ways, to the fashioning and the reception of science.

All in all, the different chapters in this volume have in common the idea that Shakespeare incorporates science and technology into his work in a dynamic and creative way, thereby drawing on the scientific controversies of his time, whether in the field of optics, alchemy, atomism, time technology or astrology, while simultaneously reinterpreting them at the dramatic level and contributing to shaping the ongoing intellectual discussion. That Shakespeare was interested, and possibly involved, in some of the stormiest scientific debates of his time is a conclusion that can certainly be derived from this collection of essays. But what they also suggest is that Shakespeare seemed to favour a balanced and rather cautious stance when he dealt with images derived from the worlds of science, superstition and technology, poised as he was between naturalism and supernaturalism, the new science and more traditional epistemologies, revolutionary systems of thought and more conventional beliefs. Finally, in the field of science as in so many others, Shakespeare's art remains primarily characterised by its ambivalence and its plurality.

Part I

Popular Beliefs

The 'Science' of Astrology in Shakespeare's Sonnets, *Romeo and Juliet* and *King Lear*

François Laroque

In *Richard II*, Act 2, scene 4, part of the king's army disbands when hearing false news:

> Welsh Captain: 'Tis thought the king is dead. We will not stay.
> The bay trees in our country are all withered,
> And meteors fright the fixèd stars of heaven.
> The pale-faced moon looks bloody on the earth,
> And lean-looked prophets whisper fearful change. (2.4.7–11)

Stars in Shakespeare are often uncanny and the harbingers of disaster, whether it be issues of war, of political succession or of love. 'I believe they are portentous things / Unto the climate that they point upon', Casca says to Cicero in *Julius Caesar* (1.3.31–2). And, as Horatio remarks in *Hamlet* Q2:

> A little ere the mightiest Julius fell,
> The graves stood tenantless, and the sheeted dead
> Did squeak and gibber in the Roman streets
> At stars with trains of fire, and dews of blood,
> Disasters in the sun; (1.1)[1]

But Roman superstition and medieval belief in the powers of celestial prophecy remained attached to ominous as well as numinous

---

[1] See Hibbard 1994: 355 (Appendix A).

moments of divine revelation. Not so astrology in early modern Europe, since it was linked to astronomy, mathematics and medicine. Indeed, what was then called the 'science' of astrology was traditionally divided into two parts, i.e. natural and judicial astrology. Natural astrology was concerned with general planetary influences, such as those on nature and agriculture or those on the human body in the medical field. Judicial astrology consisted in relatively precise predictions, derived from the respective position and influence of the planets, that were intended as advice to individuals: a *nativity*, or horoscope, was based on the map of the heavens at the time of a person's birth; *elections* denoted the process of choosing the most propitious moment, i.e. when the planets were most favourable, such as for a king's or queen's coronation, for example; finally there were *horary* questions when the astrologer tried to resolve personal problems according to the state of the heavens when the question was being posed.[2] The close links between medicine and astrology, and then between astrology and mathematics, contributed to make 'serious' astrologers such as John Dee reputed figures that were invited at the courts of kings. Moreover, the availability of 'cheap printed handbooks and guides that taught the rudiments of casting horoscopes and nativities' (Eamon 2014: 149) contributed to the popularity of astronomy in the Elizabethan era.

This being said, one must admit that the topic was a controversial one and that the feelings and opinions about astrology were quite ambivalent in those days (Chen-Morris 2016: 259–60). Undeniably, '[b]y the mid-17th century [. . .] astrology could claim no leading scientific figures' (ibid.) as had previously been the case. It is no wonder, then, that the poet George Herbert wrote, in *Jacula Prudentum or, Outlandish Proverbs, Sentences etc* (1640), that 'the manners of infants are moulded more by the example of parents, than by stars at their nativities' (Herbert 1841: 184). Yet for others the situation was quite the opposite. Indeed, here is what could still be read in a late eighteenth-century book by the English physician Ebenezer Sibly:

---

[2] On judicial astrology, see Walsham 1999: 23: 'Rooted in Babylonian learning and augmented by the Arabs and Greeks, by the late Middle Ages a loosely Christianized version of this ancient system of thought enjoyed extensive appeal and prestige: heavenly bodies were declared to be delegated agents and instruments of the deity.' Unsurprisingly, the craze for judicial astrology 'alarmed' Protestant ministers (ibid.).

a child, the moment it draws breath, becomes time-smitten by the face
of heaven, and receives an impression from all parts of heaven, and the
stars therein, which taking rise from the ascendant, sun and moon [. . .]
operate as the impressions stand and point out as with the finger of God,
the causes whence the fate and fortune of the new-born infant proceed
[. . .] (Sibly 1784: vol. 1, 17)

Before him, the Royal Society Fellow John Aubrey believed that 'we
are governed by the planets, as wheels and weights move the hands
of a clock' (quoted in Bobrick 2005: 230) and, in his *Brief Lives*, he
was careful to note the exact nativities of his characters when they
were known.

The revival and success of astrology among the elite in the second
half of the sixteenth century in England was due to the rise of mathematics, and, indeed, brilliant mathematicians like John Dee, John Allen
and Thomas Digges were also famous astrologers (John Dee cast horoscopes for both Queen Elizabeth I and the German emperor in Prague,
Rudolf II). Besides its elegant aesthetics and symbolism, one of the reasons for the intellectual appeal of astrology was the fact that it provided
people with a coherent and comprehensive system of thought. It also
had the further advantage of offering possible gains in self-knowledge
and, through this, a hope of better control and greater freedom. Simultaneously, the early modern age was a period when almanacs were
extremely popular, so that natural astrology was also at its peak, as the
many successive editions of shepherds' calendars show.

As far as the theatre is concerned, we know that the roofs of most
public houses were then decorated on the underside, as in today's
London Globe, with painted zodiacal signs and stars. Interestingly,
the physician and astrologer Simon Forman (1552–1611) described
his visits to three of Shakespeare's plays (*Macbeth, Cymbeline, The
Winter's Tale*) in his manuscript *Bocke of Plaies*, which forms part of
his manuscript collection known as *Simon Forman's Diary* (Kassell
2005: passim). As to Shakespeare's references to astrology and to
the stars, they are as numerous as they are contradictory and, in this
respect, the playwright certainly reflects the ambivalent feelings and
general perplexity of his contemporaries. This is why it seems absurd,
like in the case of Jean Richer, for instance (Richer 1990: passim), to
give astrological readings of his plays trying to make systematic connections or correspondences between the dramatis personae and the
pre-established types of the zodiac.

It is probably more rewarding to analyse the specific astrological
imagery and symbolism in order to see how this enabled Shakespeare

to understand and enlighten the relationships between man and his spatial, as well as temporal, environment. More than simple superstitions, or 'archaic' remnants, ominous signs from the planets thus become for him a means of dramatic foreshadowing, and also a way of enlarging the domestic or historical horizon of certain plays, in order to give them a cosmic and universal dimension.

The 'Prophetic Soul' of the Sonnets

In Sonnet 14, the poet or speaker poses as a mock-astrologer, referring to the various provinces of astrology, to its body of knowledge, and specific techniques, as a conceit to explore the young man's heart and to develop the idea that it is urgent for him to get married and beget children:

> Not from the stars do I my judgement pluck,
> And yet methinks I have astronomy,
> But not to tell of good or evil luck,
> Of plagues, of dearths, or seasons' quality;
> Nor can I fortune to brief minutes tell,
> Pointing to each his thunder, rain and wind,
> Or say with princes if it shall go well
> By oft predict that I in heaven find.
> But from thine eyes my knowledge I derive,
> And, constant stars, in them I read such art
> As truth and beauty shall together thrive
> If from thyself to store thou wouldst convert:
> Or else of thee this I prognosticate,
> Thy end is truth's and beauty's doom and date. (Sonnet 14)

The first two quatrains present a series of denials that the poet may be regarded as a divinity, able to read 'good or evil luck', as well as his refusal to see his verse reduced to a simple shepherd's almanac, as in Spenser's *The Shepheardes Calender* (1579), where each month is associated with a specific agricultural labour, with a constellation and sign of the zodiac to match. His function does not consist in making weather forecasts or in casting horoscopes for the great, like the court astrologer John Dee. His own type of astrology consists in reading the eyes of the young man like two fixed stars and, rather than prognosticate the death of truth and beauty if he does not get married, the poet yet hopes that his friend will soon commit himself to 'store' and beget children.

In the next sonnet, the poet 'unmetaphors' astrology in order to see it as the great active principle that rules the ways of the world (judicial astrology) and the cycles of natural growth and decay (natural astrology):

> When I consider every thing that grows
> Holds in perfection but a little moment;
> That this huge stage presenteth nought but shows,
> Whereon the stars in secret influence comment;
> When I perceive that men as plants increase,
> Cheerèd and checked even by the selfsame sky,
> Vaunt in their youthful sap, at height decrease,
> And wear their brave state out of memory: (Sonnet 15, ll. 1–8)

As Michael Schoenfeldt explains, in this context, 'at height decrease' (l. 7) suggests 'the waning of the moon (taking *at height* figuratively to mean "fullness"), the descent of the sun (taking *at height* literally, and *decrease* to mean the decline of daylight from noon to sunset), and a tidelike ebbing of once youthful sap' (Schoenfeldt 2010: 24). There is here an interesting combination of astrology and theatre that presents astral 'influence' (literally the way stars were believed to work their effects on men and women by pouring down, as in an influx, or flood, an ethereal fluid called 'influence') as spectators commenting on a play in such ways as may remain unheard by the actors. This is a brilliant variation on the age-old cliché of *totus mundus agit histrionem* (the whole world plays a part) applied to the stage of the world placed under the 'influence' of the stars that seem to govern the affairs of the world and the laws of natural growth and decay.

Finally, in the enigmatic Sonnet 107, the poet denies the pessimism of the 'sad augurs' who had predicted chaotic moments at the turn of the century, around the death of Elizabeth, England's heirless queen, here referred to as 'the mortal moon' in an allusion to one of her favourite denominations as Cynthia:

> Not mine own fears, nor the prophetic soul
> Of the wide world, dreaming of things to come,
> Can yet the lease of my true love control,
> Supposed as forfeit to a confined doom.
> The mortal moon has her eclipse endured
> And the sad augurs mock their own presages.
> Incertainties now crown themselves assured,
> And peace proclaims olives of endless age. (Sonnet 107)

So, in the sonnets, astrology is used in various ways, as a metaphor, as a form of serious art or science, but also in such a way as to cloud some sort of political message connoting impatience with the old queen and joy at the news of James I's accession.

'Star-Crossed Lovers': The World of *Romeo and Juliet*

Shakespeare's first love tragedy is characterised by its embedded sonnets that serve as references to the Petrarchan tradition and model as well as a means to formalise the cult of love in a new type of prayer, where the saint and the pilgrim share the quatrains and lines of the closing couplet in order to express the beauty and the intensity of mutual adoration. The prologue itself has been given the form of a sonnet and it serves the double purpose of exposition and forewarning in its packed, almost formulaic, lines that sound like some sort of ominous prophecy:

> Two households both alike in dignity,
> In fair Verona where we lay our scene,
> From ancient grudge break to new mutiny,
> Where civil blood makes civil hands unclean.
> From forth the fatal loins of these two foes
> A pair of star-crossed lovers take their life,
> Whose misadventured piteous overthrows
> Doth with their death bury their parents' strife. (Prologue, ll. 1–8)

It is clear that the feud between the two households in Verona comes from an 'ancient grudge' which is also marked by astrological connotations. The nativity of the two lovers, which, right from the beginning, is described as 'star-crossed', indeed seems to echo the 'fatal' loins of the two foes and shows that their passionate pilgrimage, or, as the prologue puts it, 'the fearful passage of their death-marked love' (l. 9), is bound to become a dance of death.

Interestingly, like star-gazers and almanac-makers, Shakespeare, early on in the play, calls attention to the festive background and traditions as well as to their places within their temporal context, inside the calendar. Instead of situating the Capulet feast at Christmas, as in his main source, Arthur Brooke's 1562 *Romeus and Juliet*, he shifted it to the mid-July period and replaced the allusions to Blind Fortune with repeated mentions of the role taken by the stars in his play. By doing so, he was able to associate his theme of passionate love with the period of the 'dog days', when the star known as Sirius, or Canis

Major, was predominant. Thus the fire of desire reflects the fire in the sky just as it is intensified by it, and this is constantly reiterated to the spectator by the recurrent use of the word 'rage', which, more than choler, is evocative of the dog and of the dog days, as well as allusions to 'hot days', 'mad blood' and flashes of 'lightning'. The poetic images of the play, like the many bawdy puns and double entendres (as in the servants' mock-defiant and obscene dialogue in 1.1), introduce correspondences between the heat of the sun and the brightness of the stars, on the one hand, and the concomitant fury, violence, sexual excitement and special combination of hate and love, on the other.

Indeed, according to Mercutio, noon heat seems responsible for sexual arousing ('the bawdy hand of the dial is now on the prick of noon'; 2.3.104–5), while for Friar Laurence immoderate love leads to a form of inevitable, almost mechanical destruction, or explosion:

These violent delights have violent ends,
And in their triumph die like fire and powder,
Which as they kiss consume. (2.5.9–11)

Before he meets Juliet at the feast, Romeo insists on his own misgivings and on the fateful role he ascribes to the stars:

my mind misgives
Some consequence yet hanging in the stars
Shall bitterly begin his fearful date
With this night's revels, and expire the term
Of a despised life, closed in my breast,
By some vile forfeit of untimely death. (1.4.106–11)

When he hears of Juliet's 'death' while exiled in Mantua, he exclaims, 'Then I defy you, stars' (5.1.24) before he decides to 'shake the yoke of inauspicious stars / From this world-wearied flesh' (5.3.111–12) and drinks the apothecary's poison.

But the most significant example of this astral determinism is to be found in the intriguing homophony between the name of the month and the name of the heroine, between July and Juliet, which reminds us that, after England's refusal to adopt the Gregorian reform of the calendar (Gregory, incidentally, is the name of one of the Capulets' servants), a reform introduced in all of Europe in 1582, we are still presented here with references to the old Julian, i.e. Julius Caesar's, calendar in the play. Indeed, Capulet's only surviving child owes her name to the fact that she was born on the last day of July, the 31st, on the eve of Lammastide, as the Nurse puts it. When she is asked to

give Juliet's exact age, the Nurse repeatedly quibbles on the specific
acoustics of Juliet's name, which she parses as 'Jule'/'Ay':

> 'Thou wilt fall backward when thou hast more wit,
> Wilt thou not, *Jule*?' And, by my halidom,
> The pretty wretch left crying and said '*Ay*'.
> To see now how a jest shall come about!
> I warrant an I should live a thousand years,
> I never should forget it. 'Wilt thou not, *Jule*?' quoth he,
> And, pretty fool, it stinted and said '*Ay*'.
> [. . .]
> 'Wilt thou not, *Jule*?' it stinted and said '*Ay*'. (1.3.44–59, my emphasis)

As to Romeo, who has come to the feast in order to see the fair, yet
cruel, Rosaline, he exclaims as soon as he catches sight of Juliet,
whose name he naturally does not know:

> O, she doth teach the torches to burn bright!
> It seems she hangs upon the cheek of night
> As a rich jewel in an Ethiope's ear – (1.5.43–5)

The words Jule/jewel/July are here aligned against a dark, nightly
background that sets off the bright torches and the rich jewel against
the 'cheek of night' and the 'Ethiope's ear', and this blazon creates
an anthropomorphic vision to which the bright/night rhyme adds
a chiasmic touch. It is a miniature encapsulation that combines an
acoustic name game (Jewel = Jule) with the day/dark, bright/night
environment, which marks the specific visual rhythm of the play.
'What's in a name?' Juliet asks (2.1.85). In the text of the play at
least, we see that the month of July serves as a marker of identity
associated with specific sounds and images, as any sign, which consists of the association of a signified and a signifier, but one which is
also situated in time and even inside the calendar.

As 'jewel', Juliet becomes 'starified', or rather her eyes become
stars. She transcends the division between human life and cosmic life,
between the sub- and the trans-lunary world:

> Two of the fairest stars in all the heaven,
> Having some business, do entreat her eyes
> To twinkle in their spheres till they return.
> What if her eyes were there, they in her head? –
> The brightness of her cheek would shame those stars
> As daylight doth a lamp; her eye in heaven

Would through the airy region stream so bright
That birds would sing and think it were not night. (2.1.57–64)

The eyes/stars chiasmus is both a Petrarchan cliché and an astrological allusion grounded in the idea that the heavenly bodies are made of the same elements as human bodies. So, the pathetic fallacy here indirectly justifies the subliminal astrological discourse. What's more, it foreshadows the wedding night (3.5) and the lovers' exchanges about the lark and the nightingale, the day and the night birds:

Juliet: Yon light is not daylight; I know it, I.
It is some meteor that the sun exhaled
To be this night a torchbearer,
And light thee on thy way to Mantua. (3.5.12–15)

To return to Juliet's nativity, the fact that the Nurse does not situate Juliet's birthday on 31 July but on 'Lammas Eve at night' (3.1.23), in the manner of the old Celtic calendar which divided the year into halves of winter and summer, is a way of suggesting that Juliet was probably conceived ('from forth the fatal loins of these two foes'; Prologue, l. 5) during the night of Hallowe'en, exactly nine months before. This kind of 'retrograde computation', as Rabelais calls it,[3] was both a popular and a legal 'art of memory', an empirical and fairly simple way, often used in those days, to assess whether the child was to be considered as legitimate or not.[4] So, in this nativity which associates the contrary signs of life and death, day and night, summer and winter, we find an inscription, or rather 'impression', as astrologers put it, of Juliet's destiny.

What's more, the specific traditions and games associated with the celebration of Hallowe'en are far from indifferent or insignificant in the specific astrological context of *Romeo and Juliet*. Shakespeare makes use of this principally to indicate, quite early on in the play, that a tragic mechanism with the power to abolish the frontiers between love and death may be at work. Indeed, the traditions associated with Hallowe'en were marked by a deep ambivalence. On the one hand, there were ritual games and initiations into love, through which girls could hope to glimpse the face of their future husbands;

---

[3] The expression is used by Rabelais in his *Quart Livre* (*Fourth Book*). For more on this, see Laroque 1995: 24 and Laroque 1991: 205.
[4] The tragedy of *King Lear*, for instance, opens on a discussion of the difference between Gloucester's bastard and legitimate children.

on the other hand, the customs of Hallowe'en also revived ancient Saxon beliefs that the dead returned to the world of the living during the time before the great annual festival of the dead, which was linked to the ancestor-cult of pagan times. To a public still prone to superstition and used to playing the game of calendrical associations, the fact that Juliet might have been conceived on that very night constituted a covert hint as to her destiny.

So, Shakespeare constructed his first love tragedy by weaving multiple and contrary visions of time into its dramatic text, since the seasonal and festive cycles are being short-circuited by the brief, sonnet-like moment of desire and love which also demands instant as well as intense satisfaction. During the period of the dog days, in mid-July, Sirius was the predominant star and was commonly represented as a dog with a flaming torch in its mouth, and this fire imagery is reinforced in the play by Juliet's allusion to Phaeton in 3.2.3. Indeed, if one is to believe Ovid's *Metamorphoses*, when the son of Merops goes and sees his real father, the sun god Apollo, he asks him for a sign of acknowledgement and begs the favour of driving the chariot of the sun. Seeing that he cannot refuse, Apollo warns him against the dangers attached to the task. Arthur Golding translates his passage as follows:

> By blind by-ways and ugly shapes of monsters must thou go.
> And though thou knew the way so well as that thou could not stray,
> Between the dreadful Bull's sharp horns yet must thou make thy way;
> Against the cruel bow the which th'Aemonian Archer draws;
> Against the ramping Lion armed with greedy teeth and paws;
> Against the Scorpion stretching far his fell and venomed claws;
> And eke the Crab that casteth forth his crooked clees awry,
> Not in such sort as th'other doth, and yet as dreadfully.
> Again, thou neither hast the power nor yet the skill, I know,
> My lusty coursers for to guide, that from their nostrils throw
> And from their mouths the fiery breath that breedeth in their breast.
> [. . .] My Phaeton, thou dost crave
> Instead of honour ever a scourge and punishment for to have.
> (Golding 2002: 64–5; Book II, ll. 105–34)

Lion, Bull, Archer, Scorpion and Crab are all signs of the zodiac, so that one might go as far as to say that, through the Phaeton allusion, *Romeo and Juliet*, placed under the 'secret influence' of the zodiacal canopy of the theatre's roof, can also be read as an astrological tragedy that moralises rashness, impatience, violence and haste. It takes place both in the Italian city of Verona and in the sky. Indeed, when Paris says to Friar Laurence, 'For Venus smiles not in a house of tears' (4.1.8), Venus is here both goddess and planet, while the word 'house' may be understood in the double sense of 'family' and of 'celestial section' or 'planetary division'. Here 'house of tears' means, besides the bereaved Capulet household, an inauspicious section of the heavens, perhaps the eighth house or 'house of death'. In Book II of *The Faerie Queene*, Spenser's line 'When oblique *Saturne* sate in the house of agonyes' (Spenser 2013: II, ix, 59) shows that the image was familiar to the Elizabethans, and here it adds weight to the lovers' 'yoke of inauspicious stars' (Levenson 2000: 301, n. 8).

In the Renaissance, the Phaeton myth was associated with the canicular myths, as Rabelais's *Pantagruel* shows (Rabelais 1994: 223, chap. 2). Ovid's fable was the first to express fears about growing heat and climate change, fears that were often associated with apocalyptic visions of a world destroyed by fire. Indeed, one of the marginal notes of the Phaeton episodes refers to the following quotation as 'the burning of the world' (Golding 2002: 69):

> The restless horses of the sun, began to neigh so high
> With flaming breath that all the heavens might hear them perfectly.
> [. . .] they left the beaten way
> And, taking bridle in their teeth, began to run astray.
> [. . .]
> Then wheresoever Phaëton did chance to cast his view,
> The world was all on flaring fire. The breath the which he drew
> Came smoking from his scalding mouth as from a seething pot;
> His chariot also under him began to wax red-hot.
> He would no lenger dure the sparks and cinders flying out.
> Again, the culm and smouldering smoke did wrap him round about,
> The pitchy darkness of the which so wholly had him hent
> As that he wist not where he was nor yet which way he went.
> The wingèd horses forcibly did draw him where they would.
> The Ethiopians at that time (as men for truth uphold),
> The blood by force of that same heat drawn to the outer part
> And there adust from that time forth, became so black and swart.
>   (Golding 2002: 66–9; Book II, ll. 203–301)

This astrological myth, which provides its own explanation for the apparition of black skin, is indeed a necessary complement and context to the astrological imagery in *Romeo and Juliet* which compresses it within a series of Elizabethan miniatures. Hence Jule/Juliet who 'hangs upon the cheek of night / As a rich jewel in an Ethiope's ear' (1.5.44–5) is to be read simultaneously as an erotic blazon, an astrological emblem and a sign of *discordia concors*. Poetic imagery is here particularly terse as it is condensed inside this liminal and numinous motto: *Nomen est omen.* Names will tell. 'What's in a name?' Juliet asks (2.1.85). Well, nothing and everything. As in Paracelsus's *Volumen Paramirum* (1520), which refers to what the philosopher-physician calls '*ens*', or star,[5] life is set against an ever-present yet invisible background which is not a place, a thing, or nothing, but a condition of being which keeps evading us while we may never fully grasp it as such.

Disasters in *King Lear*

The exposition scene of *King Lear* reveals Gloucester in a saucy, jolly mood but showing his rather uncouth disregard for his bastard son Edmund, while Act 1, scene 2 presents a reversal of this initial situation, due to Edmund's manipulation of his father to make him angry against Edgar with a forged letter. Now the 'old lecher' appears as a comic astrologer:

> Gloucester: These late eclipses in the sun and moon portend no good to us [. . .] nature finds itself scourged by the sequent effects. Love cools, friendship falls off, brothers divide; in cities, mutinies; in countries, discord; in palaces, treason [. . .] This villain of mine comes under the prediction: there's son against father [. . .] all ruinous disorders follow us disquietly to our graves. (1.2.101–15)

Gloucester thus expresses his pessimism regarding the three astrological levels of family, state and cosmos. He describes a chaotic situation and a world turned upside down as a result of the 'late eclipses of the sun and moon', while Edmund mocks his father's superstition and credulousness:

> Edmund: This is the excellent foppery of the world: that when we are sick in fortune [. . .] we make guilty of our disasters the sun, the moon, and stars, as if we were villains on necessity, fools by heavenly compulsion, knaves, thieves, and treachers by spherical predominance, drunkards, liars, and adulterers by an enforced obedience of planetary influence [. . .] My father compounded with my mother under the Dragon's tail and my nativity was under Ursa Major, so that it follows I am rough and lecherous. Fut! I should have been that I am had the maidenliest star in the firmament twinkled on my bastardizing. (1.2.119–30)

---

[5] See Crone 2004: 84: '[E]ach ens is one of five approaches to the same disease, a concept that emphasizes a multicausal approach to illness. In the Parenthesis the five kinds of entia are defined. They are the Ens Astrale, Veneni, Naturale, Spirituale and Dei.'

Edmund notably mocks his father for relying on the influential writings of Claudius Ptolemy, who, in his *Tetrabiblos*, explains that 'the time of conception is as important to know as the time of birth' (Rusche 1969: 161). Interestingly enough, the Egyptian astronomer supplies all the necessary elements to understand Gloucester's foolish reasoning. Indeed, as Harry Rusche makes clear, in Ptolemy's treatise

> the Dragon's Tail, the descending node where the moon crosses the ecliptic moving south, was maleficent; it detracted from a good horoscope and made a bad one worse. Thus, at Edmund's conception – his 'bastardizing' – any configuration of planets, good or bad, was affected adversely by the sinister influence of the Dragon's Tail. (Rusche 1969: 162–3)

It is no wonder, then, that, for his father, Edmund's life is bound to be disastrous. The word 'disaster' etymologically refers to a mishap resulting from the influence of an ill or inauspicious planet, like the 'stars with trains of fire and dews of blood' in *Hamlet* (1.1.116). Contrary to *Romeo and Juliet*, where the influence of the dog days, or Canis Major, and the questions of date, calendar and nativity, as well as conception, are presented as crucial elements that determine the tragic events of the play, the bastard's view corresponds to the cynical naturalist or materialistic stance. So this allows Edmund to scoff at the way or time when '[his] father compounded with [his] mother under the Dragon's tail and [his] nativity was under Ursa Major'. Like Coriolanus, Edmund dreams that a man may be 'author of himself' (*Coriolanus* 5.3.36), that his success or misfortune may be entirely due to him alone rather than to some smiling star or favourable Fortune.

The opposition between Gloucester and Edmund here reflects the Renaissance controversy over the validity of astrology, especially of judicial astrology. This questioning of the old 'science' of astrology was not totally new at the time as it was found in Puritan writings such as those of William Fulke and Philip Stubbes. Their polemic treatises, respectively the *Antiprognostication* (1560) and *The Anatomie of the Abuses in England* (1583), contain the following observations:

> Sycknesse and healthe depende upon dyvers causes, but nothyng at al upon the courses of the stares (Fulke 1560: sig. Dvii)

> it is the [. . .] wickednesse of our owne harts, that draweth us to evill [. . .] and not the starres, or planets (Stubbes 1882: vol. 2, 56–8)

On the other hand, Gloucester's insistence on the portentousness of eclipses was a view that was often professed, and it was also aimed at countering naturalistic positions. So the opening of *King Lear* presents multiple mirrors and perspectives for viewing its action. The Renaissance dilemma is here poised between two rival cosmic explanations: supernaturalism in the astrological determinism of Gloucester, and naturalism in Edmund's materialistic views.

Conclusion: Shakespeare's Ambivalence

In the three specific examples of the sonnets, *Romeo and Juliet* and *King Lear*, we have seen that astrology serves very different and sometimes contradictory purposes, and that it corresponds to sometimes divergent images. In the example from the sonnets, astrology plays the function of an analogical language, of a metaphoric system meant to strengthen the links between poetry and prophecy in the speaker's repeated attempts to read the complex signs leading to the young man's heart. In *Romeo and Juliet*, astrology is used as a substitute for the machinery of Fate, insofar as a double correspondence is established between Juliet/jewel and the stars in an enlarged Petrarchan conceit, on the one hand, and between the disaster of Phaeton and the tragedy of the 'star-crossed lovers', on the other. The issues of nativity, calendar combinations or computing, and saints' days suggest that the characters' humours are totally dependent on the positions of the stars during the dog days in Verona. So, in the play, astrology does seem to provide some sort of overall explanation as to the ill luck of the title characters as well as for the various accidents and fortune reversals. In *King Lear*, father and bastard son take up opposite stances as they try to explain the current disorders of the kingdom, but of course the spectator knows that Gloucester is blind and manipulated by his evil son (just as when he is indeed blind after his eyes are pulled out, he will be guided by his good son Edgar). In a tragic and cruel world, in which the gods are absent or silent, astrology becomes just one of the ways to try and understand evil and the mystery of things, and thus more of a religion, or superstition, than a real science.

Whether Shakespeare uses astrology in such contrastive ways to deliberately exploit its rich symbolism as well as its dire forewarnings, be they true or deceptive, for stylistic and structural reasons, or whether he, up to a certain extent, believed himself in the ominous power of its dark – and, as we today know, totally unscientific – predictions, is a point that would certainly deserve further research.

Chapter 2

Staging Devils and Witches: Had Shakespeare Read Reginald Scot's *The Discoverie of Witchcraft?*

Pierre Kapitaniak

European demonological reflections and writings gathered momentum towards the end of the fifteenth century boosted by the success of *Malleus maleficarum* (1486). They endured until the eighteenth century and, as Stuart Clark has shown (Clark 1997), remained perfectly compatible with the new empirical ideas advocated by the members of the Royal Society in the late seventeenth century. Although it never achieved the institutional status of a recognised discipline of knowledge, the scientific preoccupation with ghosts, devils and witches fuelled the theories developed by theologians, jurists and physicians, as suggested by the scientific pretensions of the neologisms chosen as titles for some of such treatises, like Sébastien Michaelis's *Pneumalogie* (1587) or Noël Taillepied's *Psichologie* (1588). The status of demonology as science is in itself questionable: the term 'science' usually refers to the object of its study rather than to the intellectual activity of thinking and writing about the devil and his minions. Indeed, it is magic, alchemy, astrology, divination and necromancy that are systematically labelled as science, in the traditional meaning of 'knowledge or understanding acquired by study' (*OED* 2). Yet the proceedings of demonological investigations often correspond to a more modern definition of science, as those scholars deal with 'a connected body of demonstrated truths or with observed facts systematically classified and more or less comprehended by general laws, and incorporating trustworthy methods [. . .] for the discovery of new truth in its own domain' (*OED* 4b). It is a peculiarity of England that the first serious treatise devoted to the science of demons – Reginald Scot's *The Discoverie of Witchcraft* (1584) – was a most radically sceptical one, and therefore aimed at undermining the legitimacy of the field itself, though it did not prevent a profuse literature from thriving for nearly another century.

In his recent monograph about Martin Del Rio, Jan Machielsen insightfully remarks that Renaissance demonology must be considered much more as textual scholarship than experimental science (Machielsen 2015: 264), for it was above all grounded in traditional use of exempla retained for their authority through time. It is precisely such a vision of demonology, and therefore demonology itself, that Scot attacked in *The Discoverie of Witchcraft*. And it is precisely such an attitude to textual authorities that Scot opposed when he repeatedly insisted that 'truth must not be measured by time: for euerie old opinion is not sound' (Scot 1584: sig. B3r).

Despite legends about King James I ordering publicly to burn Reginald Scot's *The Discoverie of Witchcraft*, the book was an ongoing success from the moment it was published, more often meeting with approval than with condemnation. Among those who approved of Scot's ideas and who plundered them eagerly were several generations of London playwrights. In *The Discoverie of Witchcraft* they found ample inspiration for all their supernatural figures that became so successful on Elizabethan and Jacobean stages, and one can only wonder whether the slow evolution from genuinely supernatural ghosts, demons, witches and wizards towards more and more intangible figures or mere tricks (Kapitaniak 2008: 684–7) is not to be accounted for on the basis of Scot's successful influence on London theatre-makers and theatre-goers.

Over the centuries, scholars have ascribed various words, images and themes found in Shakespeare's plays to his reading of *The Discoverie*, and the aim of the present chapter is to reassess Scot's genuine influence on Shakespeare's drama throughout his career by systematically confronting the diverse intertextual evidence. This intertextual investigation will in turn inform the question of Shakespeare's recourse to demonology as a science. But in order to shed new light on this debt it is necessary to examine the wider impact of Scot's book on the circle of London dramatists of the Elizabethan and Jacobean periods.

A *Grimoire* for Playwrights

Quite logically, the first one to delve into Scot's cornucopia of incantations and superstitions was his Kentish fellow countryman John Lyly. In *Gallathea* (1585 or 1588), Lyly devoted the subplot to the adventures of three brothers cast ashore after a shipwreck (*Gallathea*, 2.3., in Lyly 2000). Rafe, one of the brothers, finds himself alone in the woods and witnesses a dance of fairies, immediately followed by the arrival of Peter, the Alchemist's apprentice, soon joined by his master. In this long scene Peter, who is fed up with serving his deceitful master, uses Rafe's enthusiasm to present him as a new apprentice, so that he himself can run away. The whole scene is interspersed with alchemistic vocabulary that is drawn from two classic literary sources on the subject – Chaucer's *Canon's Yeoman's Tale* and Erasmus's *Alcumista* – that Lyly shares with Reginald Scot. Most critics and editors of the play count Scot among Lyly's direct sources but there is actually little evidence that Lyly was using *The Discoverie* rather than the original texts mentioned above. There are even some details that indicate that Lyly was using Chaucer's original, since the four spirits are enumerated in the correct order (1. Quicksilver; 2. Orpiment), while Scot reverses the first and the second spirit. Perhaps one might see a residual echo of his reading of *The Discoverie* in the choice of the 'Spanish needle' that Rafe would see turned into a 'silver steeple' (3.3.13–14) since Scot uses the 'Spanish needle' in one of his legerdemain tricks (Scot 1584: 327). In fact, what mainly pleads for Lyly's knowledge of Scot's book is that both men were from Kent and shared very sceptical views on supernatural phenomena, not to mention the intriguing fact that they are nowadays the main sources of our knowledge about Mother Bombie, a cunning woman who seems to have become famous for her magic tricks in Rochester, and to whom Lyly devoted a play a few years later.

There are other playwrights for whom the identification of Scot's influence is much more unequivocal. The three examples that I will examine here have in common the borrowing of incantations (either fragments or whole) from *The Discoverie* as if from a *grimoire*. In William Percy's *Arabia sitiens, or, A Dreame of a Drye Yeare: a Tragaecomodye*, known nowadays as *Mahomet and His Heaven* (1601), we find Geber, 'an old Figure caster', whom Percy transforms from an accomplished alchemist into a (quite inefficient) master of black magic, having a familiar named Smolkin at his service (Percy 2006: 42). When in Act 3 Geber is cunningly dispossessed of his magic ring and purse, he conjures his familiar together with a long list of devils borrowed straight from Scot's translation of Johann Weyer's *Pseudomonarchia dæmonum* (1577):

> Raum, Haiphas, Foculor, Brifrons, Num
> Gamagin, Zagun, Urias, Valuc,
> Smolkin, Furcas, Murmur, Caim, Vapalour,
> Come along, helpe your Master, I adjure. (3.7.3–6)

The different names are not reproduced in the original order,[1] but still retain clusters of subsequent names, probably reordered for the sake of metrical requirements. 'Num' seems to be a misreading of 'Vine'; the combination of 'V' followed by 'i' might have been amalgamated into a single letter. The spellings with a preference for the 'i' over the 'y' strongly suggest the use of Scot's version rather than Weyer's, and of course the addition of 'Smolkin', which is the name of Geber's familiar in the play, explains this last difference from the source. William Percy was the brother of the famous Henry Percy, 9th Earl of Northumberland, remembered as the 'Wizard Earl' due to his alchemical experiments. It is Henry's vast library that provided William with a copy of *The Discoverie* (Percy 2006: 53).[2]

A similar kind of borrowing, though much more voluminous, appears in the comedy of *The Two Merry Milkmaids* (1620), written by one 'J.C.', whom some critics have identified as the actor John Cumber. In the comedy, Bernard, a Wittenberg student and apprentice to the magician Landoffe, tries to conjure a devil in the absence of his master and uses incantations from his *grimoire* that once again come straight from Scot's translation of Weyer. I have shown elsewhere the numerous parallels and borrowings in the play (Kapitaniak 2007: 211–17), and the fact that Scot, rather than Weyer, is the playwright's direct source is further suggested by a plain allusion in the short preface 'The Printer to the Reader', which plays with the two words of Scot's title: 'If there be discouerie made of the Coniuring Words, you'le find the Witchcraft' (J.C. 1620, n.p.)·

Ben Jonson disguised a similar allusion to *The Discoverie* in his *Masque of Queenes* (1609), as demonstrated by W. Todd Furniss. Throughout the masque Jonson shows off his scientific erudition by inserting overwhelming marginalia providing references to numerous classic and state-of-the-art demonologists, among whom are Sprenger and Institoris, Spina, Psellos, Bodin, Rémy, Del Rio and many others, but neither Johann Weyer nor Reginald Scot. Furniss identified at least one passage for which Scot's influence may be established beyond doubt (Furniss 1954: 346–7).[3] He explains the

---

[1] The order both in Scot and in Weyer is as follows: Zagan, Orias, Valac (Weyer: Volac), [. . .] Vapula.

[2] Incidentally, it is interesting to note that Henry had in his library a copy of Weyer's *De Præstigiis dæmonum* (1583), which he heavily annotated (Batho 1960: 254).

[3] The passage in Scot (Scot 1584: 10) provides Jonson with 'egg-shells' and 'Awger holes' (Furniss 1954: 357).

absence of clear references by the fact that Jonson may have wanted to avoid naming the two authors whom King James had openly condemned in the preface to his *Dæmonologie* (1597).

But the most impressive appropriation of *The Discoverie*, in terms of both quality and quantity, is the way Thomas Middleton builds the characters of Hecate and her sister-witches in *The Witch* (1615). Where other playwrights pick up occasional quotations from a page or two of the treatise, Middleton displays a much deeper knowledge of its contents: it is obvious that he read it through, as he borrows from nine book divisions out of sixteen (Nicholson 1886: 550–2; Middleton 2012: 358–60). While we can imagine that Percy copied the passage he later used in his play from his brother's book (which might further account for the spelling mistakes), Middleton must have worked with *The Discoverie* virtually on his desk, delving into it whenever he needed a picturesque detail. He even copies from Scot nine lines in Latin from Ovid's *Metamorphoses* to enhance one of Hecate's angry speeches to the disbelieving Duchess (5.2.20–8).

One might even suppose that it was the reading of *The Discoverie* that inspired Middleton with the title of his *Changeling*, as the OED attributes the first occurrence of the specific meaning of the word as 'a child [. . .] supposed to have been left by fairies in exchange for one stolen' to Reginald Scot.

This quick survey establishes that *The Discoverie* was a well-known and well-read book among the circle of London playwrights. If so many of them used Scot to document and build their supernatural characters and effects, it is only fair to assume that Shakespeare did so too.

The Case of *Macbeth*: Shakespeare or Middleton?

It is the tragedy of *Macbeth* that usually springs to mind when one looks for Scot's influence on Shakespeare. The play is haunted with both witches and ghosts, yet critics and editors of the play have long suggested that the apparition of Banquo's ghost at the banquet is closest to an account found in Pierre Le Loyer's *Treatise of Spectres*, recently translated into English by Zachary Jones (Muir 1977: 216; Paul 1950: 57–9; Clark and Mason 2015: 96–7). But even in the case of witches, the study of the sources is complicated by the fact that the surviving text of *Macbeth* is the version found in the 1623 Folio, which had in the meantime been revised for the stage by Thomas Middleton around 1616. The most visible traces of Scot's text in

Macbeth occur in two scenes involving witchcraft in the tragedy: 3.5 and 4.1. Recently Sandra Clark and Pamela Mason have offered a clear synthesis of the arguments concerning Middleton's possible revision of the play (Clark and Mason 2015: 323–36) and my purpose, in this chapter, is not to resume the discussion of such likelihood. As far as Scot's potential echoes are concerned, the case is less problematic since the two songs were imported from Middleton's own play *The Witch*, while there seems to be something of a consensus about Middleton's hand in the whole of Act 3, scene 5. Consequently, the evident presence of textual traces of *The Discoverie* is inconclusive if we try to assess Shakespeare's knowledge of the source.

We can hear the difference in the playwrights' indebtedness to Scot through a comparison of two passages: the first, a song inserted in Middleton's *The Witch*, and the second, the cauldron song in *Macbeth*, unanimously accepted to be Shakespeare's.

> Titty and Tiffin, Suckin
> And Pidgin, Liard and Robin!
> White spirits, black spirits, grey spirits, red spirits!
> Devil-toad, devil-ram, devil-cat, and devil-dam!
> (*The Witch*, 1.2.1–4; inserted at 4.1.43)

On closer examination, the beginning of this song turns out to be borrowed word for word from a passage in the appended *Discourse upon divells and spirits*: 'Now, how *Brian Darcies* he spirits and shee spirits, Tittie and Tittin, Suckin and Pidgin, Liard and Robin, &c: his white spirits and blacke spirits, graie spirits and red spirits, diuell tode and diuell lambe, diuells cat and diuells dam' (Scot 1584: 542). Middleton merely rearranges Scot's sentence into verse, suppressing a syllable here and there for the sake of the rhythm and replacing one single word with another that rhymes equally.

Let us now turn to a stanza from the famous cauldron song that the witches sing in the same scene:

> Fillet of a fenny snake
> In the cauldron boil and bake.
> Eye of newt and toe of frog,
> Wool of bat and tongue of dog,
> Adder's fork and blindworm's sting,
> Lizard's leg and howlet's wing,
> For a charm of powerful trouble,
> Like a hell-broth boil and bubble. (4.1.12–19)

Scot of course provides many descriptions of charms and potions in his book, but the closest to the ingredients listed by Shakespeare does not bear any convincing resemblance: 'the haire growing in the nethermost part of a woolues taile, a woolues yard, a little fish called Remora, the braine of a cat, of a newt, or of a lizzard: the bone of a greene frog, the flesh thereof being consumed with pismers or ants' (Scot 1584: 124). Not only are there few precise ingredients in common, but some of those chosen by Shakespeare (like 'bat', 'blindworme' or 'howlet') never occur in *The Discoverie* at all. Moreover, the quoted passage is in fact translated by Scot from Weyer's *De Præstigiis dæmonum*, and given its faint resemblance to Shakespeare's song, both Scot's translation and Weyer's original might be no more than vague recollections of a distant reading. What makes Scot's influence even less likely in this case is the fact that the very term 'cauldron' is used only once in the whole treatise (Scot usually preferring 'pot' or 'vessel'), when he describes a recipe of the witches' ointment: 'after buriall steale them out of their graues, and seeth them in a caldron, vntill their flesh be made potable' (Scot 1584: 41).[4] It is therefore impossible to postulate Scot's influence here despite a few resemblances in the choice of ingredients because those mentioned could have been found in many other texts dealing with witches' charms.

Nevertheless, there are other echoes that might appear more convincing, either because of closer thematic proximity or because of the infrequency of some words in the English language. In Act 1, scene 3, the witches evoke their power to create winds ('I give thee a wind'; 1.3.11), which is also one of the first faculties that Scot mentions about witches:

> Such faithlesse people (I saie) are also persuaded, that neither haile nor snowe, thunder nor lightening, raine nor tempestuous winds come from the heauens at the commandment of God: but are raised by the cunning and power of witches and coniurers; insomuch as a clap of thunder, or a gale of wind is no sooner heard, but either they run to ring bels, or crie out to burne witches. (Scot 1584: 2)

Yet Scot is just one source in which Shakespeare could have read about this particular power, and it is revealing that most critics of

---

[4] It is worth mentioning that this passage too is a translation, this time from the *Malleus maleficarum*.

the play cite other authors for the idea, Thomas Nashe being the most frequent (Muir 1977: 12; Brooke 1990: 101; Clark and Mason 2015: 138). Further into the same scene, Banquo's reaction when the title of the Thane of Cawdor is first applied to Macbeth takes up a demonological commonplace – 'can the devil speak true?' (1.3.108) – further expanded upon a few lines later: 'The instruments of darkness tell us truths, / Win us with honest trifles, to betray's / In deepest consequence' (125–7). Muir chooses to rely on James's *Dæmonologie*, yet the quotation he bases his evidence upon is quite remote from the context as it does not refer to truth. Scot, on the other hand, does recount an anecdote about a priest who caught the devil professing the truth:

> On a time the diuell went vp into a pulpit, and there made a verie catholike sermon: but a holie preest comming to the good speed, by his holinesse perceiued that it was the diuell. So he gaue good eare vnto him, but could find no fault with his doctrine. And therefore so soone as the sermon was doone, he called the diuell vnto him, demanding the cause of his sincere preaching; who answered: Behold I speake the truth, knowing that while men be hearers of the word, and not followers, God is the more offended, and my kingdome the more inlarged. (Scot 1584: 481)

Once again, Scot is not alone in tackling the subject, and Shakespeare may equally have remembered reading Ludwig Lavater's *Of ghostes and spirites walking by nyght*, where the author devotes a chapter to how 'The diuell sometimes vttereth ye truth, that his words may haue the more credite, and that he may the more easely beguile them' (Lavater 1572: 173; Frye 1984: 20). Since no striking textual parallel may be noted between any of these works, such thematic coincidence is not sufficient to plead for a borrowing.

John Dover Wilson thought he had spotted another trace of *The Discoverie* in Macbeth's panic-stricken reaction to Banquo's ghost: 'If charnel-houses and our graves must send / Those that we bury back, our monuments / Shall be the maws of kites' (3.4.68–70). Wilson quoted Scot's opinion on Nebuchadnezzar: 'I am not ignorant that some write, that after the death of Nabuchadnez-zar, his sonne Eilumorodath gaue his bodie to the rauens to be deuoured, least afterwards his father should arise from death, who of a beast became a man againe' (Scot 1584: 102; Clark and Mason 2015: 222). Yet, as above, the context is much too remote and, even worse, Scot himself is borrowing here from the English translation of Agrippa's *Of the Vanitie and uncertaintie of Arts*.

At the end of the same scene, once the guests have departed, Macbeth resumes his desperate broodings:

Stones have been known to move, and trees to speak;
Augures, and understood relations, have
By maggot-pies and choughs and rooks brought forth
The secret'st man of blood. (3.4.121–4)

Muir, as well as Clark and Mason, provides two quotations from Scot in which the phenomenon of augury is described in similar terms:

These cousening oracles, or rather oraclers vsed (I saie) to exercise their feats and to doo their miracles most commonly in maids, in beasts, in images, in dens, in cloisters, in darke holes, in trees, in churches or churchyards, &c: where preests, moonks, and friers had laid their plots, and made their confederacies a forehand, to beguile the world, to gaine monie, and to adde credit to their profession. This practise began in the ekes of *Dodona*, in the which was a wood, the trees thereof (they saie) could speake. And this was doone by a knaue in a hollowe tree, that seemed sound vnto the simple people. (Scot 1584: 165)

They add a second passage, which uses another example to make the same point:

Diuine auguries were such, as men were made beleeue were done miraculouslie, as when dogs spake; as at the expulsion of *Tarquinius* out of his kingdome; or when trees spake, as before the death of *Cæsar*. (Scot 1584: 208)[5]

In fact, the evidence becomes even more convincing when one carries on to the next chapter of *The Discoverie*, where Scot also mentions the rooks: 'Naturall augurie is a physicall or philosophicall obseruation [. . .] as by the crieng of rooks' (Scot 1584: 208). Yet, as often with Scot, it must not be forgotten that the first quotation is but a translation of Weyer's *De Præstigiis dæmonum*, while the other two are 'Englished' from Agostino Nifo's *De Auguriis*. Surprisingly, editors of the play delve into James's *Dæmonologie* to document the last

---

[5] See Muir 1977: 96–7. Though they do not acknowledge the debt, Clark and Mason cite the same passages, mistaking '9.18' for '11.18' undoubtedly because Muir was using Roman figures (226).

line of Macbeth's speech with his opinion on cruentation (the fact that a corpse will bleed in the presence of the murderer), forgetting that, long before James, Scot had already commented on a similar phenomenon in *The Discoverie* and, despite his general scepticism, allowed for its reality (Scot 1584: 303–4; Muir 1977: 97; Clark and Mason 2015: 227).

There are a few more parallels that invite an intertextual relation with Scot because of the rarity of the vocabulary. One such case is the use by Donalbain of the expression 'an auger hole' (2.3.123). Stephen Greenblatt was the first to spot the echo of a passage from *The Discoverie*: 'They [i.e. witches] can go in and out at awger holes, and saile in an egge shell, a cockle or muscle shell, through and vnder the tempestuous seas' (Scot 1584: 10; Greenblatt 1993: 125; Clark and Mason 2015: 195). Yet, though rare, Shakespeare may have noted the same expression in Arthur Dent's *The plaine mans pathway to heauen* (1601) – 'creepe into an auger hole' – where the spelling matched Shakespeare's more closely.[6]

Another lexical parallel occurs in Act 4, scene 1, when Macbeth addresses the witches and their supernatural powers:

> I conjure you by that which you profess
> (Howe'er you come to know it), answer me.
> Though you untie the winds and let them fight
> Against the churches, though the yeasty waves
> Confound and swallow navigation up,
> Though bladed corn be lodged and trees blown down,
> Though castles topple on their warders' heads,
> Though palaces and pyramids do slope
> Their heads to their foundations, though the treasure
> Of nature's editorial germen tumble altogether
> Even till destruction sicken, answer me
> To what I ask you. (4.1.49–60)

Again critics call for Scot's authority on the power over winds and remark on the lexical resemblance between 'bladed corn' and 'corne in the blade':

> And first *Ouid* affirmeth, that they can raise and suppresse lightening and thunder, raine and haile, clouds and winds, tempests and earthquakes.

---

[6] The case is much more evident for Ben Jonson, who uses the same expression in his *Masque of Queenes*.

> Others doo write, that they can pull downe the moone and the starres. Some write that with wishing they can send needles into the liuers of their enimies. Some that they can transferre corne in the blade from one place to another. (Scot 1584: 10)[7]

If we are to assess Scot's influence here on the basis of the whole speech, the result is negative for, apart from the expression 'bladed corn' (which is far from similar, by the way), Scot's description is even further from Shakespeare's lines than, say, the famous lines from Ovid which openly inspire Scot in this passage, and which he later quotes and translates.

The last such echo is found towards the end of the scene, when Macbeth comments to Lennox on the Weird Sisters that have just vanished: 'Infected be the air whereon they ride, / And damned all those that trust them!' (4.1.137–8).[8] Besides their faculty of flying which repeatedly occurs in *The Discoverie*, as in many works on witches, it is their power to vitiate the air that justifies the parallel with a passage from Scot's opening chapter: 'The Imperiall lawe (saith Brentius) condemneth them to death that trouble and infect the aire: but I affirme (saith he) that it is neither in the power of witch nor diuell so to doo, but in God onelie' (Scot 1584: 3). As the quotation shows, Scot is not the first to remark on that power, and actually the quotation is not second- but third-hand, as Scot translates once again one of his main sources, Weyer's *De Præstigiis dæmonum*.

All in all, each of these parallels is subject to caution, as it is either textually too remote from the source, too general in its topic to be pinned down to a single source, or, in many cases, when the similarity is more evident, Scot's source is mediated. Yet the accumulation of clues repeatedly pointing towards *The Discoverie* deserves in itself further investigation into the other plays in Shakespeare's canon.

The Case of *A Midsummer Night's Dream*: Bottom and Puck

Another famous supernatural element begs comparison with Scot's treatise. In *A Midsummer Night's Dream*, Bottom is magically transformed into an ass by the playful intervention of Oberon's

---

[7] See Muir 1977: 109; Clark and Mason 2015: 238.
[8] See Clark and Mason 2015: 246.

servant Puck (3.1.91ff). The story of man's metamorphosis into
an ass was commonplace, well known through the accounts of
Ovid and Apuleius, and in the more specific field of demonological inquiries it featured in the *Malleus maleficarum* (1486) and
Jean Bodin's *Démonomanie des sorciers* (1580). These also happen to be among the main sources of Scot's *Discoverie* (Scot 1584:
169–70), which, in addition, inserts another variation of the transformation borrowed from Giambattista della Porta's *Natural
Magic* (1558) (Scot 1584: 347). So far Shakespeare would have
been spoilt for choice, and given his extensive use of Ovid, Scot
would not be his first. However, Marion Gibson has noted an echo
in *The Comedy of Errors* that seems much more reminiscent of
Scot's translation of Bodin's tale of the sailor changed into an ass
and forced to work as a beast of burden for the wicked witch.
When Dromio of Syracuse expresses his fear of Ephesian witches
he envisages the following scenario: 'I amaz'd ran from her as a
witch. And I think [. . .] she had transform'd me into a curtal dog,
and made me turn i'th'wheel' (3.2.136–41). Gibson argues that the
'reference to a human being made to serve a witch as a domestic
animal is the closest Shakespeare comes to repeating Scot's story
in full' (Gibson 2014: 189). As in the case of *Macbeth*, Scot's tale,
being second-hand, weakens the argument of his influence on
Shakespeare.

Yet Bottom is not the only character in *A Midsummer Night's
Dream* that seems to leap out from the pages of *The Discoverie*. It
is true that Robin Goodfellow is a recurrent example in Scot's witty
criticisms of popular belief, and Harold F. Brooks has provided a
long list of relevant passages from Scot (Brooks 1979: 146–9). Still,
Peter Holland has somehow qualified this filiation (Holland 1994:
35), because the only indubitable link between the two texts is the
aggregation of terms that designate Oberon's servant:

> Either I mistake your shape and making quite,
> Or else you are that shrewd and knavish sprite
> Called Robin Goodfellow. Are not you he
> That frights the maidens of the villagery,
> Skim milk, and sometimes labour in the quern
> And bootless make the breathless huswife churn,
> And sometime make the drink to bear no barm,
> Mislead night wanderers, laughing at their harm?
> Those that 'hobgoblin' call you, and 'sweet puck',
> You do their work, and they shall have good luck. (2.1.32–41)

The three alternative names for the character are here 'Robin Good-fellow', 'sweet puck' and 'hobgoblin', and they are very close to three names among the long list of 'bugbears' and superstitions that maids taught children according to Scot: 'Robin good fellowe, the spoorne, the mare, the man in the oke, the hell waine, the fierdrake, the puckle, Tom thombe, hob gobblin, Tom tumbler, boneles, and such other bugs' (Scot 1584: 153). The only notable variation concerns the pair 'Puck'/'Puckle'. But even in this case there is a possible rival source. In *A Treatise of Treasons against Q. Elizabeth* (1572), John Leslie evokes the same superstition and retains the same three names: 'that spirit or pook that we call Robin Goodfellow, or Hobgoblin' (Leslie 1572: f.5r). Shakespeare's 'Puck' differs as much from Leslie's 'Pook' as from Scot's 'Puckle', which suggests that the three authors may have drawn their inspiration from local folklore.[9] However, in this case, other elements point to Scot rather than any other writer or oral source. The fairy's description of Puck takes up a description that Scot gives in his preface. The presence of the word 'churn', whose introduction in English the *OED* ascribes to Scot,[10] confirms a cluster of words, images and themes that pass from Scot to Shake-speare and that are present in the very passage I have conjured up when discussing *Macbeth* for the rare occurrence of the expression 'auger holes'. All this makes it worth quoting the passage in full:

> They can raise spirits (as others affirme) drie vp springs, turne the course of running waters, inhibit the sunne, and staie both day and night, changing the one into the other. They can go in and out at awger holes, and saile in an egge shell, a cockle or muscle shell, through and vnder the

---

[9] Schleiner pleads for Shakespeare's greater originality without totally discarding Scot's parallel but also Spenser's *Epithalamion* (1595), which mentions 'Pouke' and 'hob Goblins' (Schleiner 1985: 65–8), while Gibson favours a direct debt to Scot (Gibson 2014: 158–89). Robin Goodfellow and hobgoblins are also mentioned in *Tarlton's newes out of Purgatorie* (1590: 2): 'although thou see me heere in the likenes of a spirite, yet thinke me to be one of those *Familiares Lares* that were rather pleasantly disposed then indued with any hurtfull influence, as *Hob Thrust, Robin Goodfellow* and such like spirites (as they terme them of the buttry), famozed in every old wives Chronicle for their mad merrie pranckes. Therefore sith my appearance to thee is in resemblance of a spirite, thinke that I am as pleasant a goblin as the rest, and will make thee as merry before I part, as ever *Robin Goodfellow* made the country wenches at their Cream boules.' See also Wall 2001: 70–1.

[10] *OED*: '*churn*: b. intr. To work a churn, make butter.' For this sense the first two occurrences given by the *OED* are Scot and Shakespeare.

tempestuous seas. They can go inuisible, and deprive men of their pri-
uities, and otherwise of the act and vse of venerie. They can bring soules
out of the graues. They can teare snakes in peeces with words, and with
looks kill lambes. But in this case a man may saie, that *Miranda canunt
sed non credenda Poetæ*. They can also bring to passe, that chearne as
long as you list, your butter will not come; especiallie, if either the maids
haue eaten up the creame; or the goodwife haue sold the butter before in
the market. Whereof I haue had some triall, although there may be true
and naturall causes to hinder the common course thereof: as for example. Put a little sope or sugar into your chearne of creame, and there will
neuer come anie butter, chearne as long as you list. (Scot 1584: 10–11)

The various words and thematic echoes from this passage sound
relevant to the treatment of fairies in *A Midsummer Night's Dream*,
and seem slightly more present there than they were in *Macbeth*.
Moreover, unlike many excerpts quoted before, which were often
themselves quotations copied or translated by Scot from other
works, here we have one of the author's original reflections. Most
significantly, Scot opposes here (as he does throughout his treatise)
various superstitious beliefs to a series of rational explanations
meant to discredit the former once and for all. And it is such a structural divide, drawing a clear line between reality and fantasy, that
also seems to lie at the heart of Shakespeare's conception of the two
worlds in *A Midsummer Night's Dream*. But before drawing conclu-
sions from these observations, let us search for further resonances
in the plays.

The 1590s: Demons and Anecdotes

The 1590s offer a few more echoes that may be traced back to Scot's
Discoverie of Witchcraft. In *The Merry Wives of Windsor*, Ford, who
doubts his wife's fidelity, evokes several names of demons that are
intended to pale in comparison next to 'cuckold': 'Terms, names!
"Amaimon" sounds well, "Lucifer" well, "Barbason" well; yet they
are devils' additions, the names of fiends' (2.2.297–9).

Marion Gibson signals that Shakespeare had already used
'Amaimon' in *1 Henry IV* (2.4.317) and was later to recycle 'Barbason'
in *Henry V* (2.1.49) (Gibson 2014: 25–6). If Lucifer is too widely used
to be of any pertinence, the other two can be traced to the long list that
Scot translates from Weyer's *Pseudomonarchia dæmonum*. Among
those names we find a demon called Marbas or Barbas, immediately
followed by Amon or Aamon, as well as, a little further on, another one

called Amaimon or Amaymon. It is likely that Shakespeare confused or deliberately amalgamated the two. To explain the change from Barbas to Barbason, J. H. Walter argued that Shakespeare fused the demon Barbas with the name of a French gentleman, Barbason, that he may have come upon when reading *Holinshed's Chronicles* (Walter 1954: 33). Thomas Wallace Craik, on the other hand, suggested that Shakespeare simply added a syllable to Barbas to obtain a trisyllabic name for the sake of the metre (Craik 1990: 138).

One would logically expect similarly to trace the demon in *2 Henry VI* to the same source. Yet no Asmath is found in *The Discoverie*, or an approximate variation of it. Neither can the usual Latin formula of invocation – '*Conjuro te*' (1.4.21) – be traced back to Scot, as the latter systematically translates it into English ('I conjure thee . . .'). Yet even the trilogy of *Henry VI* offers something to build our evidence of Scot's influence, which has not been noticed before. In *1 Henry VI* Charles embellishes his praise of Joan of Arc with a series of commonplaces:

> Was Mohammed inspirèd with a dove?
> Thou with an eagle art inspirèd then.
> Helen, the mother of great Constantine,
> Nor yet Saint Philip's daughters were like thee. (1.3.119–22)

These lines have received considerable attention over the years, not only because of the dense cluster of commonplaces packed up in those four lines, but also because the passage became one of the cornerstones of the debate over the authorship of the first act of *1 Henry VI*. It is not the place here to develop the particulars of the debate between the proponents of Shakespeare as the sole author (Cairncross 1960: passim; Harlow 1965: passim) and those defending collaboration with another or even several other dramatists. As far as our intertextual quest is concerned, suffice it to say that among the candidates for collaboration, the most serious case is that of Thomas Nashe, whose authorship of the first act seems now to have been accepted by a majority of editors and critics (Taylor 1995; Burns 2000; Vickers 2007). Charles's ironic encomium of the Pucelle betrays a thematic proximity to two other texts: Henry Howard's *A defensative against the poison of supposed prophesies* (1583) and Thomas Nashe's *The Terrors of the Night* (1594). The most extensive analysis of the evidence can be found in Harlow's article, where the texts were discussed to confirm Shakespeare's authorship, but I propose to review the relevant passages in order to bring another text into play.

In *A defensative against the poison of supposed prophesies*, a book that was no longer a novelty by the early 1590s, the earl of Northampton alludes in passing to Mahomet's dove in the following terms: 'Mahomet, the glosing sycophant, was inspired with a Dove' (Howard 1583: sig. C3v; Harlow 1965: 275). Michael Hattaway provides Howard as the source for the commonplace (Hattaway 1990: 82) but the mere fact that the allusion occurs here is of course not enough, as the story was well known, notably thanks to *The Golden Legend*.[11] What makes *A defensative* a credible source is the fact that it also provides two other elements from Charles's speech. Later in the book, a longer excerpt combines Saint Philip's daughters with Helen:

> The like simplicity was vsed by Simeon, Anna, Zachary, Elizabeth, Agæbus, and the daughters of S. Philipp. Thus Constantine the first, and as I thinke it may be sayde without offence: the best Christian Emperour that euer was, (for in him and Helene, began first that prophecy of Esay to take place) (Howard 1583: sig. Ll4v)[12]

There is even an allusion to an eagle linked with prophecy that editors seem to have overlooked, preferring to identify Shakespeare's eagle with the symbol of Saint John the Evangelist (Hattaway 1990: 82; Burns 2000: 140). Writing about the strange omens recounted by Plutarch in the *Life of Dion*, Howard expounds upon the significance of the eagle in the following terms:

> Three straunge accidents fell out at one time in Siracuse, as we reade in Plutarche. First a Pigge was brought into the worlde without eares [. . .] Secondly, the salte water on a sodaine became fresh [. . .] Thyrdly, an Eagle chaunced to snatch a Partisane out of a Souldiers hand [. . .] Vppon the thirde I would imagine that as an Eagle is Iouis ales, the byrd of Iupiter, and therefore kinge of all the rest: so might it signifie, that a greater Prince then he that ruled Siracuse, should make an end of warre, and depriue the state of the strongest weapon. (Howard 1583: sig. D4v)

---

[11] It was Line Cottegnies who remarked in her French translation and edition of the play that this anecdote was already found in *The Golden Legend* (Cottegnies 2008: 1,456). See Caxton, *The Golden Legend*, 'the historie of Saint Pelagyen the Pope', f.CCCCixr.

[12] The allusion to Saint Philip's daughters is even repeated in two other places: 'S. Philippes daughters, were inspyred wyth the like gift, and (as Eusebius reports) renouncing all vaine pleasures of the worlde, remained vergins whyle they lyved' (sig. Pp4v); 'Dyd euer Olda the Prophetesse, the blessed virgin Elyzabeth, or saint Phillips daughters, paint theyr faces, colour their heaires, weare garish apparell, practise vsurie, or play at the tables' (sig. Qq1v).

If this is indeed the eagle that Shakespeare has in mind, the association of the four commonplaces makes a strong case for Howard's direct influence.[13] The question also remains as to who was the reader of *A defensative*. Nashe, who shows an acute interest in the science of devils and apparitions, seems to have assimilated Howard's book, as we find the same series of echoes in his *Terrors of the Night, or A Discourse of Apparitions*, published in 1594 in the form of a mock-demonological treatise. Editors usually quote Nashe on 'the chast daughters of Saint Philip' (Nashe 1594: sig. G3v; Burns 2000: 140) and Nashe also refers to 'the Doue where with the Turkes hold Mahomet their Prophet to bee inspired' (Nashe 1594: sig. B4v–C1r). The combination of these elements clearly suggests knowledge of Howard's text that is often used as an argument to establish Nashe's authorship of the passage. Yet it is also Nashe who establishes a link with Reginald Scot, when we put the above quotation back in its context:

> Socrates Genius was one of this stampe, and the Doue where with the Turkes hold Mahomet their Prophet to bee inspired. What their names are, and vnder whome they are gouerned, the Discouerie of witchcraft hath amplified at large, wherefore I am exempted from that labour. (Nashe 1594: sig. B4v–C1r)

This suggests that Nashe draws his reference not so much from Howard as from Scot, a clue that editors have left unnoticed for many decades. Yet when one looks at the way the Mahomet commonplace is presented in *The Discoverie*, the link with Scot appears indisputable:

> Euen as another sort of witching priests called *Aruspices*, prophesied victorie to *Alexander*, because an eagle lighted on his head: which eagle might (I beleeue) be cooped or caged with *Mahomets* doue, that picked peason out of his eare. (Scot 1584: 171)[14]

---

[13] Even without the help of this passage, Harlow deemed the other three elements as sufficient to conclude that the author of *1 Henry VI* must have read Howard (Harlow 1965: 276).

[14] Once again, as for *Macbeth*, the case for Scot's influence is slightly weakened by the fact that Scot translated the whole sentence from Weyer. Surprisingly enough, although Edward Burns calls for the authority of Reginald Scot to comment more generally on the links between the figure of Joan of Arc and the context of witch hunts in England, he never thinks of establishing an intertextual link between the two works (Burns 2000: 48).

Before Shakespeare, Scot combines the image of the dove with that of the eagle, which neither Howard nor Nashe did. The eagle in this case refers to Alexander the Great and we may infer that Shakespeare's does likewise, even though, as we have seen, editors draw a parallel with the eagle of Saint John the Evangelist.[15] Scot may also have provided the allusion to 'S. Helen, mother of king Constantine' (Scot 1584: 260),[16] even though we find no trace of Saint Philip's daughters in *The Discoverie*. Once we know which eagle to look for, we may wonder if Nashe's text does not bear a remote echo of this anecdote too, be it in a very indirect way:

> Could any man set downe certaine rules of expounding of Dreames, and that their rules were generall, holding in all as well as in some, I would beginne a litle to list to them, but commonly that which is portentiue in a King is but a friuolous fancie in a beggar, and let him dreame of Angels, Eagles, Lyons, Griffons, Dragons neuer so, all the augurie vnder heauen will not allot him so much as a good almes.
>   Some will obiect vnto mee for the certainety of Dreames, the Dreames of Cyrus, Cambyses, Pompey, Caesar, Darius, & Alexander. For those I answer, that they were rather visions than Dreames, extraordinarily sent from heauen to foreshew the translation of Monarchies. (Nashe 1594: sig. D4r)

If Scot appears as a more likely source for the first part of Charles's speech, Howard still provides a more convincing one for the other two lines. We know that Nashe had read both Howard's and Scot's books because he explicitly quotes the two works in *The Apologie of Pierce Pennilesse* (1592):

> in that far-fetched sense may the famous *defensative against supposed Prophecies*; and the *Discovery of Witchcraft* be called notorious Diabolical discourses [. . .] for they also entreat of the illusions and sundry operations of spirits. (Nashe 1592: sig. H4v)

The evidence suggests that the author of Charles's speech – whether Shakespeare or Nashe – read both sources, and that he borrowed

---

[15] Incidentally, Scot's intertext and the identification of the eagle offer more coherence to Charles's list of similes that was felt problematic by some critics. See Michael Taylor: 'This pagan climax to Charles's list of Joan's spiritual ancestors jars with the saintly Christian ones' (Taylor 2003: 117).

[16] Also translated from Weyer.

the association between Alexander's eagle and Mahomet's dove from *The Discoverie*, and the juxtaposition of Constantine's mother and Saint Philip's daughters from *A defensative*.

Even if Shakespeare did not write those lines and cannot be held responsible for the choice of commonplaces there, his collaboration with Nashe may have enabled him to hear of Scot's book and to read it. This might explain the presence of likely echoes of *The Discoverie* in later plays, analysed above, to which we might add the allusion to a certain 'Monarcho' that appears in Boyet's description of Armado in *Love's Labour's Lost*:

> This Armado is a Spaniard that keeps here in court,
> A phantasime, a Monarcho, and one that makes sport
> To the Prince and his bookmates. (4.1.97–9)

As we saw on several occasions, this may as easily have come from Scot ('The Italian, whom we called here in England, the *Monarch*, was possessed with the like spirit or conceipt'; Scot 1584: 54) as from Nashe's *Have with you to Saffron-Walden* ('*Monarcha* the Italian, that ware crownes on his shooes'; Nashe 1596: sig. M2v), or from any of the few texts that alluded to it in the 1590s.[17]

Conclusion: The Extent of Scot's Influence

The long list of possible parallels that I have examined in this chapter yields no easy conclusion. The great majority of the evidence is weakened either by the existence of rival sources, which Shakespeare might equally have read, or by authorship problems that muddle the analysis and no longer necessarily inform us on what Shakespeare may have read. Maybe it is this uncertainty that accounts for Marion Gibson's ambivalent position in her recent *Shakespeare's Demonology* concerning the playwright's use of *The Discoverie*. Within the individual entries of her dictionary she states quite strongly that Scot was 'one of Shakespeare's favourite demonological writers' and 'one

---

[17] In his edition of *Love's Labour's Lost*, H. R. Woudhuysen refers only to Nashe (Woudhuysen 1998: 180), while William C. Carroll acknowledges both Scot and Nashe in his (Carroll 2009). For a list of the main sources on the subject, see Halliwell-Phillipps 1855: vol. 4, 326.

of Shakespeare's more evident sources for demonological material' (Gibson 2014: 62, 116), while in the general introduction she implies a distinction between two kinds of material by admitting that '[i]t is *likely* that Shakespeare knew these works, and *certain* that Harsnett was an important source for [*King Lear*]' (Gibson 2014: 4, my emphasis).

My investigation of Scot's influence on Shakespeare has adopted reverse chronological order, from *Macbeth* to *1 Henry VI*, dictated by the bulk of possible borrowings in each of the examined plays. Indeed, if, partly thanks to Middleton, the text of *Macbeth* seems to be teeming with echoes, *1 Henry VI* contains only one anecdotal comparison that may be traced to *The Discoverie*. What's more, taken individually, none of those parallels is really conclusive, but it is the very accumulation of such echoes that pleads in favour of Shakespeare's knowledge of the book. If we analyse *Macbeth* such as it must have been written in 1606, before Middleton's revisions, Scot's impact rather consists of remote echoes than precise borrowings. Back in 1595, *A Midsummer Night's Dream* shows a denser network of words and images passing from the treatise on to the stage. Finally, the history plays and a few early comedies display punctual words or anecdotes, in all likelihood because the subject is further away from Scot's main preoccupation, which makes *The Discoverie* only a peripheral source in the genesis of those plays.

The hypothesis that I would like to venture at the outcome of this intertextual exploration is that Shakespeare may well have read the sceptical treatise early in his career, influenced by his collaboration with Thomas Nashe on *1 Henry VI*, and did not feel the need to come back to the text later, as Middleton did, content to rely on his more or less vague memories as time went by. Such a chronology would also account for the lack of demonic names from *The Discoverie* in *2 Henry VI*, written before *1 Henry VI* – that is, before Shakespeare learned about Scot's book from Nashe.

Among other demonological sources used by Shakespeare we may count the famous *Of ghostes and spirites walking by nyght* (1572) by Ludwig Lavater, for which there are fewer exact textual parallels than for Scot; the already quoted *Treatise of Spectres* (1605) by Pierre Le Loyer, which merely yielded the anecdote of a banquet ghost for *Macbeth*; and *A Declaration of Egregious Popish Impostures* (1603) by Samuel Harsnett, which provided Scot with a few names of devils. Unlike Ben Jonson or Thomas Nashe, who take a learned interest in demonology as a field of knowledge, Shakespeare treats demonology as no more than a sourcebook of anecdotal folklore; he never really takes up or develops any demonological theories on the moral nature of apparitions or on the implication of the demonic pact or the use of *maleficium*. Far from considering demonology as a science whose findings might feed his reflection on human nature, his is a purely literary interest, perfectly illustrating what Marianne Closson described as the 'intermediary space of belief which favoured the emergence of fantastic literature' (Closson 2000: 378), even within the field of demonology itself.

Part II

Healing and Improving

'Remedies for Life': Curing *Hysterica Passio* in Shakespeare's *Othello*, *Macbeth* and *The Winter's Tale*

Sélima Lejri

I thought good to make known the doctrine of this disease so far forth as may be in a vulgar tongue conveniently disclosed, to the end that the unlearned and rash conceits of divers might be thereby brought to better understanding and moderation, who are apt to make everything a supernatural work which they do not understand [. . .] who are ready to draw forth their wooden dagger if they do but see a maid or woman suffering one of these fits, conjuring and exorcising them as if they were possessed with evil spirits.

Edward Jorden, *A Briefe Discourse of a Disease Called the Suffocation of the Mother*, 1603, 'The Epistle Dedicatorie'

It gives me an estate of seven years' health, in which time I will make a lip at the physician. The most sovereign prescription in Galen is but empiricutic, and, to this preservative, of no better report than a horse-drench.

William Shakespeare, *Coriolanus*, 2.1.93–6

Acquainted with the medical theories of his time, Shakespeare has Menenius jokingly mock the much-praised empiricism that his era inherited from the Greek physicians, and more specifically from Galen. To 'make a lip' (2.1.94) at the validity of the most unchallenged and infallible authority's works that saw 590 editions between 1500 and 1600 in Europe was perceived as a sacrilegious act, and members of the Royal College of Physicians of London who dared to question them were reproved (Thomas Neely 2004: 71; Irving

1970: 156). In a period marked by the rediscovery, republication and translation of the classics, the interest in the material provided by the Greek medical canon grew intensely. In particular, the explorations in the field of gynaecology roused much attention and caused it to develop into a specialised topic of medical expertise. This was mainly attributable to the Galenic theories of humours and vapours but also to the Hippocratic treatise *Diseases of Women*, which was translated from Greek into Latin as early as 1525 (Peterson 2010: 3). Consequently, the human body and its long-hidden mysteries were uncovered from those ancient physicians under the fascinated eyes of the Elizabethan anatomists (Sawday 1995: 266).[1]

Unlike their Greek predecessors and chiefly Galen, who was 'a radical materialist' (García-Ballester 2002: 159), the Elizabethan and Jacobean physicians were nourished by Protestant theology and could hardly keep distinctions clear between religious and medical discourses as they mingled the rational and the emotional, the empirical and the metaphysical, the humoral and the spiritual. Timothy Bright and Robert Burton, to name the most famous representatives of the Elizabethan and Jacobean periods as far as medicine is concerned, were both ministers and physicians, or, as Burton puts it, 'spiritual physician[s]' (Burton 2016: 'Democritus Junior to the Reader', 29). Their understanding of melancholy, its causes and its cures rests mostly on ethical discourses of sin, punishment and repentance (Lund 2010: 57; Thomas Neely 2004: 15; Reed 1952: 68). Burton relies equally on Stoic philosophy in his attempt to find a compromise between theological and material explanations for the afflictions of the body and the soul, inseparable as they are (Lund 2010: 168–71). Moreover, these physicians wrote in a period when the number of treatises reinforcing the folklore and superstitious beliefs in demonism, witchcraft and possession was quite considerable and at a time when even those sceptical about such beliefs, whilst denouncing them as fake and fraudulent, provided as much material on the topic. Hence, from the widely circulating *Malleus maleficarum* (1486) (Levak 2004: 57–69) to the *Compendium maleficarum* (1608) (ibid., 99–104) that supported beliefs in witchcraft, and from *The Discoverie of Witchcraft* (1584) (ibid., 285–93) – to

---

[1] In his book *The Historie of Man* (1578), John Banister – reputedly the first Englishman to serve as subject of a painted anatomy lesson – attests in total worship of Hippocrates and Galen: 'What secret so dainty, that they have not uncovered? Yea, what mystery so covert, the door whereof they have not opened?' See Banister in Neill 1997: 123.

which James VI of Scotland responded through his *Dæmonologie* (1597) (ibid., 140–5) – to *A Declaration of Egregious Popish Impostures* (1603), a vast corpus was available to nourish the imagination on the demonised body, offering a more culturally rooted alternative to the rationalised and medicalised body, still imperfectly accommodated to the nascent discourses of medical therapies being imported from another culture.

It is in these intersecting discourses dividing exorcists, witch-finders and theorists from empiricists and physicians that the woman's body, and the womb in particular, is inscribed. More than any bodily part, the womb exacerbates the fanaticism of the devil-proselytisers when it escapes the knowing look of their materialist opponents. Indeed, despite the growing attention to female biology and uterine affects, and despite the development of gynaecology and embryology, little was concretely known of the womb, which remained hidden and seldom dissected. All the medical treatises of the time are based on Galen, who never practised dissection, except for bloodletting (Eccles 1982: 76).[2] Therefore, already believed to be predisposed by nature to an anatomical deficiency that would make it vulnerable to demonic assaults, the female body becomes much more submitted to the intersecting discourses of the sacred, the satanic, the superstitious and the supernatural as it presents signs of abnormal behaviour (Paster 1993: 25). The studies of female reproductive organs and uterine diseases were incorporated into a large and varied nosography of diseases, from delirium to melancholia, the latter being capacious enough as to allow subdivisions, such as in Burton's encyclopaedic work on the topic, where there is a section devoted to 'Symptoms of Maids, Nuns, and Widows' Melancholy' (Burton 2016: I, 176). It is to this particular type of melancholy that the present chapter is devoted, with reference to the pamphlet of Edward Jorden, the first physician who, at the beginning of the seventeenth century, theorised it after the Hippocratic and Galenic tradition in his contention against the interpretations of the disease by contemporary clerics (and even doctors) as bewitchment.

Although they retain a streak of the spectacular Senecanism that characterises the scenes of madness in his early works, Shakespeare's

---

[2] In her *Midwives Book* (1671), Jane Sharp asserts: 'Rarely the object of dissection, the interior of the pregnant female anatomy is hardly ever *seen*' (see Sharp in Laoutaris 2008: 10). Moreover, Laoutaris explains that anatomists relied on the rare corpses of dead women and foetuses which they could obtain from midwives and graveyards (11).

Jacobean plays incorporate the new scientific discourses of his time regarding female diseases. The relationships of the three couples in *Othello* (1604), *Macbeth* (1605) and *The Winter's Tale* (1611), as well as the pathological symptoms of the heroines in these particular plays, have much in common as far as the scientific notions of humoralism, melancholy and hysteria are concerned. Therefore, I will first introduce briefly the causes of these illnesses through a portrait of the husbands who share a comparable approach toward female biology and sexuality, before examining the consequences those apprehensions have on the three women within the medical frame of Shakespeare's sources, hence from the perspective of early modern expressions of emotions that rest mainly on Galenic naturalism. I will conclude this chapter with an analysis of the discourses of cure and remedy that the diseased women are submitted to. This analysis will therefore apprehend the phenomenon of hysteria – with its causes, symptoms and effects – from two complementary perspectives: contemporary moral treatises and medical texts, as well as modern gender criticism. The latter weaves in smoothly with the former as it offers a historical reading of the language of affect in both the scientific primary sources and the literature of the time, and excavates the early modern mental landscape with its proper conception of the human body and emotions.

'Tremor Cordis' and 'Milk of Human Kindness': Male Anxieties

In his attempt to inform Polixenes of his master's state, Camillo, in *The Winter's Tale*, stops short of explicitness: 'I cannot name the disease' (1.2.381). Iago, on the contrary, has a metaphor for it: 'the green-eyed monster' (*Othello*, 3.3.180). As they indulge in jealousy and paranoid delirium, both Leontes and Othello express sexual anxiety towards their respective wives, whom they fantasise as oversexed. Admittedly 'declined / Into the vale of years' (3.3.282–3), Othello lacks confidence in his sexual potency, preferring instead to project himself into an asexual being. He even publicly disclaims any intention '[t]o please the palate of [his] appetite / Nor to comply with heat' (1.3.262–3) in taking Desdemona with him to Cyprus, confessing that 'the young affects' are 'in [him] defunct' (1.3.263–4). In an attempt 'to cerebralize and desexualize their relationship' (Traub 1992: 37), he insists that his only wish is 'but to be free and bounteous to [Desdemona's] *mind*' (1.3.267, my emphasis), shaking off Desdemona's overtly sexual invitation and expectation: 'My *heart's* subdued / Even to the utmost *pleasure* of my lord' (1.3.252–3, my emphasis). Hence his dual castration anxiety at the prospect of failure in satisfying the one whose 'greedy ear / Devour[ed] up [his] discourse' (1.3.151–2), heralding a voracious sexual appetite, and at the undeniable fact, magnified by Iago, that young and vigorous rivals, such as Cassio, could offer a better match for her. Consequently, as Edward Snow explains, his fear of adultery mingles with 'thralldom to the demands of an unsatisfiable sexual appetite in woman' (Snow 1980: 407).

Leontes is confronted with a similar, but more acute, anxiety-breeding threat from a friend of his age and one of peculiar significance to him. Indeed, he detects his wife's overruling sexuality in the fact that she manages to detain Polixenes in their court for a longer sojourn, thereby stirring in him a dual feeling of betrayal, since he feels fooled not only by her, whom he now sees as 'too hot' (1.2.110), but also by his childhood 'playfellow' (1.2.82), who has turned out to be untrue to their friendship that bears homosexual undercurrents.[3] Interestingly enough, Camillo's explanation of the reason for Polixenes's acceptance bears a streak of irony: 'To satisfy your highness and the entreaties / Of our most gracious mistress' (1.2.234–5). The verb 'satisfy' is of course not lost on Leontes, who angrily reiterates it twice (1.2.235–6) as he now grapples with his potential failure to meet his wife's sexual 'entreaties'. Just as Othello reads Desdemona's moist hand as a testimony of her 'liberal heart' requiring 'a sequester from liberty' (*Othello*, 3.4.38–40), Leontes views Hermione's pregnancy as a token not of him being her exclusive sexual partner but of her uncontrollable corporeality that autonomously 'rounds apace' (2.1.17) and ostentatiously offers its fertile reproductive organs to other masculine claimants (Leon Alfar 2003: 167–8; Rosenfield 2002: 97).

As to Macbeth's sexual anxiety, it arises from his feelings of inadequacy in the masculine world of political power (Leon Alfar 2003: 112). Assessing her husband's virility by the degree of ruthlessness and brashness he would demonstrate to wrench royal power, Lady Macbeth deplores his natural predisposition, which she presents as that of a lactating mother, 'too full of the milk of human

---

[3] On the homosexual relationship between Polixenes and Leontes, see Nuttall 2007: 346–9 and Orgel 1989: 18.

kindness' (1.5.16),[4] and piques him with remarks that question his sense of manhood (Harding 1969: 246–7). She turns his claim to power into a sexual bargain, warning that 'From this time / Such [she] account[s] [his] love' (1.7.38–9), and she taunts him with the squeamishness of the impotent lover 'afeard / to be the same in [his] own act and valour / As [he is] in desire' (1.7.39–41) (Ramsey 1973: 288). Macbeth readily responds by successfully committing regicide and returning to her proudly brandishing his phallic-like daggers. This success earns him the unique ecstatic acclaim that acknowledges his manly worth – 'my husband?' (2.2.13) – soon superseded by 'infirm of purpose' (2.2.50), a rebuke with a strong resonance of sexual impotency when he recoils from taking back the daggers to the king's chamber. It is worth noting here that Macbeth's sexual castration is exacerbated by the thought of seeing 'the seeds of Banquo' (3.1.71) become kings. He is faced with his rival's fertility when all he has is a 'barren sceptre' (3.1.63), with no prospect of him generating progeny with a murderous wife who fantasises herself 'dash[ing] the brains out' of the 'babe that milks [her]' (1.7.55–8).

Having thus for common ground the dread of overweening or perverse female sexuality on the one hand and the need to outsmart masculine rivals on the other, Shakespeare's three tragic heroes find a comfortable psychosomatic outlet in acts of violence as a substitute for their failed sexual experience. Indeed, all three undergo the same process of displacement whereby the affect of love degenerates into its opposite and culminates in acts of violence and revenge. Stephen Greenblatt discerns such a pattern in *Othello*, in which he reads the cause of Desdemona's death not uniquely in Iago's slander but just as much in Desdemona's awakening of 'the deep currents of sexual anxiety in Othello' that makes 'his insupportable sexual experience [. . .] displaced and absorbed by the act of revenge' (Greenblatt 2005: 250). Othello pursues his 'bloody thoughts with violent pace' and exchanges 'humble love' for 'wide revenge' (3.3.460–1). His psychosomatic transformation whereby his body gives in to a spectacular epileptic fit (4.1.43–8) resembles Leontes's '*tremor cordis*', ascribed by Timothy Bright and Robert Burton to a dangerous 'adustness', or overheating of the blood (Houston Wood 2002: 188–90).

---

[4] 'Like the witches, Lady Macbeth discloses a desire to determine her destiny through the control of bodily substances; substances associated with the dangerously permeable female reproductive body. Believing Macbeth to be "too full o'th' milk of human kindness" (1.5.16), she seeks to contaminate this nurturing material' (Laoutaris 2008: 185).

Tormented by a 'humour / That presses him from sleep' (2.3.38–9), he sends Hermione to prison and harbours the soothing thought of seeing her 'given to the fire' to ensure 'a moiety of [his] rest' (2.3.8). If the latter heroes' bodies register excess of the sanguine element, Macbeth's bears the symptoms of choler or yellow bile, suggested by his description of his solitary and homicidal journey as eventually 'fall'n into the sere, the yellow leaf' (5.3.25).[5] His rather cold reception of the news of his wife's death is comparable to Othello's and Leontes's detachment from their respective wives, one which, incidentally, makes their passionate revenge on Desdemona and Hermione all the more paradoxical. In his account of the psychology of love entitled *The Blazon of Jealousy* (1615), Benedetto Varchi reads this strange detachment from the object of love as a form of melancholy generated in the jealous man.[6] In fact, such a propensity to rashness, melancholy and depression in the three tragic heroes may be read from the perspective of early modern humoralism as psychosomatic disturbances caused by excess of blood and of black and yellow bile (Paster 2004: 61–5).

'A malady most incident to maids': *Hysterica Passio*

At the sheep-shearing feast presided over by Flora, the goddess of spring and fertility, Perdita informs the shepherdesses of the causes of greensickness. She finds a ready example in the 'pale primroses / That die unmarried ere they can behold / Bright Phoebus in his strength' (4.4.122–4). Having now Florizel for a match, Perdita seizes the opportunity of the feast that celebrates lushness and fertility to implicitly warn her young unmarried friends against sexual abstinence. Describing the anaemia of the primroses as 'a malady / Most incident to maids' (4.4.124–5), Perdita voices the medical

---

[5] For a thorough exploration of the colour yellow in the medical tradition of Shakespeare's time, see Hickey 2015.

[6] 'So puissant and potent is this our desire which we have to enjoy that party which we love solely and alone, without the society and company of any other whatsoever, as that (many times) when this our high-prized commodity chanceth to light into some other merchant's hands, and that this our private enclosure proveth to be a common for others, we care no more for it, but give it altogether over, quite extinguishing and quenching in us not alone the jealousy we had of the same, but likewise the hot love and affection we bore it before.' Benedetto Varchi, *The Blazon of Jealousy*, quoted in DiGangi 2008: 208.

discourse that early modern England recovered from the Hippocratic and Galenic obstetrical writings prescribing marriage – hence therapeutic sexual intercourse – as a remedy for hysteria. Indeed, the Hippocratic authors of *Diseases of Women* explain the ailments of the womb in terms of humoral imbalance whereby dryness leads to the 'violent dislocation' of the uterus, which 'rushes and goes upward toward the moisture'. Consequently, 'when phlegm flows down from the head to the abdomen, the womb, now heavy with moisture, goes back to its place' (Hippocrates, trans. Hanson 1975: 583).

Likewise, sixteenth- and seventeenth-century physicians understood the pathology of love on the basis of gender, whereby hysteria was the terminology attributed to the feminine equivalent of masculine lovesickness, called 'amorous melancholie' by Andreas du Laurens in his *Discourse of the Preservation of Sight* (1597) (Wells 2007: 3; Thomas Neely 2004: 103). Moreover, they recognised hysteria as 'a sort of Madness', as in Riverius's description in *The Practice of Physick* (1655) (Eccles 1982: 82), or again as *furor*, as in the quasi-scientific analysis Lucretius makes of it in his influential *De rerum natura* (Lucretius 1998: IV, 1,030–191). Nevertheless, they gave it diverse names such as hysteria or *hysterica passio*, womb-fury or *furor uteri*, and even the Suffocation of the Mother, or *Suffocatio* and *Strangulatus uteri* (Gilman 1993: 12–13). The first medical treatise to mention it as 'the strangling of the wombe' is Philip Barrough's *The Method of Physicke* (1583), but it was Edward Jorden's treatise, *A Briefe Discourse of a Disease Called the Suffocation of the Mother* (1603), that used for the first time the Latin name *hysterica passio* and that dealt with it exclusively. This medical treatise relies mainly on Galen's *De causis morbus* and *De locis affectis*, which put forth the same causes and symptoms as did the Hippocratic works, with a particular focus on humours and vapours. In fact, Galen expounds that women, unlike men, do not evacuate their seed unless through sexual intercourse and that its accumulation can be converted into corrupt and venomous humour, causing the womb, or stones, to become yellowish and putrid and to send forth an odious vapour which reaches the brain and affects all the other humours of the body (Jorden 1991: chap. 2, chap. 6). Such beliefs were endorsed by several medical treatises besides Jorden's, such as Riverius's *The Practice of Physick* (1655) and Sharp's *The Midwives Book* (1671) (Eccles 1982: 77).

However, hysteria does not only affect young maids. Still referring to Galen's theses, Jorden observes that '[d]ivers women enjoying the benefit of marriage, yet through the suppression of their ordinary evacuation, fall into this disease' (Jorden 1991: chap. 6). The most relevant case is that of widows, in whom early modern physicians from Jorden to Burton detect similar symptoms. It should come as no surprise that, at the end of *The Winter's Tale*, Paulina accepts to sacrifice her own femininity and sexuality and, having long been used to widowhood, gets prepared for a lonely and mournful life upon 'some withered bough' (5.3.134) until death comes. If the loss of her mate 'that's never to be found again' (5.3.135) becomes for her the motif of death, it is because she knows the suffering a woman incurs from sexual abstinence. And if Leontes consolingly offers her Camillo as a husband, it is because he has benefited from her expertise in bringing his wife back to life, or, in medical terms, in helping his wife to recover from hysteria. Paulina has therefore gained insight into the implications of the disease thanks to Hermione's own experience of it. Indeed, being denied sexual gratification, Hermione shares, along with Desdemona and Lady Macbeth, the lot of the virgins and widows in whom physicians such as Riverius perceive symptoms of 'womb-furie': 'it may also betide married women that have impotent husbands or such as they do not much affect, whereby their Seminary Vessels are not sufficiently disburthened' (Riverius in Eccles 1982: 82).

Of the three cases, Desdemona's is the most problematic, and Calderwood's supposition regarding her sexuality is highly probable: 'Desdemona remains unbedded or incompletely bedded throughout the play' (Calderwood 1989: 125). Halfway between the 'pale primrose' and the wife deprived of 'the benefit of marriage', Desdemona sees that, as soon as she marries Othello, 'the rites for why [she] love[s] him are bereft [her]' (1.3.257). Ironically, the 'heavy interim' that she says she would endure because of 'his dear absence' (1.3.260–1), if she does not accompany him to Cyprus, is exactly what she endures despite his presence. Besides, she articulates her love as overtly sexual longing and asserts herself as 'the autonomous subject of her own desire' (Berger 2013: 204): 'My downright violence and storm of fortunes / May trumpet to the world' (1.3.249–50). Having hardly spent the wedding night with her husband, she wakes up from her slumber with a solicitous request: 'will you come to bed my lord?' (5.2.25). Lastly, it is for false charges of adultery but also for 'the loves [she] bear[s] to [him]' (5.2.43) that the Moor takes no gratification in strangling her. His gesture enacts the same effect that the womb is believed to generate in the victim of hysteria: 'most commonly, it takes them [women] with choking in the throat' (Jorden 1991: chap. 2). Othello therefore acts as a helpless physician who, instead of resuscitating to life his female patient, administers the deadly strangulation pre-scribed by his chief practitioner Iago (4.1.192) (Peterson 2010: 105). Furthermore, the convulsive fits of hysteria are usually followed by a swoon which, according to the physicians' observations, may last for some hours or even days, giving the impression that the victim is dead (Jorden 1991: chap. 3; Eccles 1982: 77). After having lain motionless for a long while, dead in the eyes of Othello and Emilia, Desdemona, with a sudden but feeble start, comes back to life and desperately attempts for the last time to urge for a proper cure: 'Commend me to my kind lord' (5.2.129). Before even committing the murder, Othello anticipates his failure to find the necessary 'Promethean heat / That can thy light relume' (5.2.12–13) and which can, in Kaara Peterson's words, 'revive Desdemona's suffocated body' (Peterson 2010: 104). Revivification after suffocation and a death-like swoon make up the typical stages of *hysterica passio*. The medical treatises of the time mention cases of hysteria where the cessation of respiration and the stiffness of the body last for such a long time that the victim is taken for dead. Edward Jorden records that 'diverse errors have been com-mitted' on hysterical women who lay dead and who 'have afterwards beene found to have life in them and have risen up in their burials' (Jorden 1991: chap. 3). Likewise, in his 1688 medical treatise, the doctor Chamberlain argues: 'Doubtless many are buried in such fits (for they last sometimes twenty four hours or more, and the bodies grow cold and rigid like dead carkasses) who would return if time were waited on and means used' (Chamberlain in Eccles 1982: 79). The best means that Chamberlain and the other practitioners pre-scribe to draw the uterus down and evacuate the overabundant seed is sexual intercourse, but the recourse to other surrogate means is also recommended, such as massages on the patient's lower abdomen or genital parts by a midwife, or even venesection (Eccles 1982: 83; Gilman 1993: 37–8). Hence, it is possible to read Desdemona's last moment, when 'she stirs again' (5.2.98), as a process of revivifica-tion that stops short of accomplishment. She stands halfway between Lady Macbeth, for whom there is no hope of reanimation from her self-inflicted strangulation, and Hermione, whose revivification is successfully carried out.

It is upon hearing terrible news that both Lady Macbeth and Hermione collapse: King Duncan's death for the former, her own son's death for the latter. Nonetheless, when considered within the larger scope of the ladies' psychological conditions, the two swoons may be inscribed in the discourse of hysteria and melancholy. Hence, Lady Macbeth's swoon may be regarded as a somatic reaction to two combined pathological symptoms: unevacuated blood and seed.[7] Urging the spirits to 'make thick [her] blood' (1.5.41), Lady Macbeth induces her own disease translated by the accumulation and corruption of blood that leads to the 'stop[ping] up' of all passages for other humours and fluids (1.5.42) (Fox 1979: 129). The word 'passage', which she employs, belongs to the medical terminology of Renaissance gynaecological texts and, as Jenijoy La Belle explains, 'it generally refers to the tract through which the blood from the uterus is discharged' (La Belle 1980: 382). In particular, Edward Jorden's medical investigation expounds that such a manifestation, called amenorrhoea or the stoppage of the terms, is proper not only to menopausal women but also to young women in the grip of 'green-sickness' (Jorden 1991: chap. 1, chap. 6; Laoutaris 2008: 187). The second symptom – unevacuated seed – is caused by Lady Macbeth's unsatisfied desire, which she seeks to appease via the perversely eroti-cised route of violence, murder and power (Paster 1993: 168). Frus-trated by such an attempt, she confesses in highly sexualised terms: 'Naught's had, all's spent, / Where our desire is got without con-tent' (3.2.6–7). Getting worse, her hysterical swoon gives way to the much more spectacular symptom of sleepwalking, attested by Robert Burton as 'troublesome sleep' amongst maids, nuns and widows suf-fering from melancholy (Burton 2016: I, 176). Again, during her somnambulistic agitations, her sexual dissatisfaction is compulsively expressed amidst her unconscious and desultory remembrances of the king's murder: 'To bed, to bed [. . .] Come / Come, come, come. Give me your hand. What's / Done cannot be undone – To bed, to bed, to bed!' (5.1. 56–8) (Levin 2002: 43).

If Lady Macbeth deplores that her desire be 'got without content' (3.2.5), Hermione laments her loss of life when her three joys are denied her: her son, her newborn baby and, above all, 'The crown and comfort of my life, your [Leontes's] favour' (3.2.92). All three sources of gratification hinge essentially upon her sexual and mater-nal functions, and the deprivation thereof causes her syncope, fol-lowed shortly by her alleged death (3.2.145, 201). Sixteen years later, she is brought back to life in an equivocal way. To Leontes, who is 'mocked with art' (5.3.68), Hermione is a perfectly true-to-life

---

[7] William Harvey names these two symptoms as parts of the preternatural diseases of the womb: 'the depraved effluxion of the Terms or the use of Venus much inter-mitted and long desired' (see his *Anatomical Exercitations* (1653), quoted in Eccles 1982: 74).

statue that is ready to be awakened to reality. Leonard Barkan sees a particular parallelism between Hermione and Michelangelo's statue of *Night*, the long-sleeping mother whose 'still sleep mocked death' (5.3.20) and whose stillness, grief and melancholy are highlighted by the anatomist Vasari (Barkan 1981: 648). However, the other protagonists know that Hermione has been kept alive in a chapel. Hence, her revivification is not the literal bringing into life of a Pygmalion-like sculpture, but the reanimation of a frozen body from a hysterical hibernation or 'numbness', the length of which being part of the plot's requirements (Peterson 2010: 143). With the assistance of Paulina, her midwife, Hermione meets a better fate than Desdemona and Lady Macbeth. Aware that she has offered her mistress only a preamble of cure, Paulina makes Leontes accept a wife of her choosing by pinning down her aim: 'She shall be such / As, walked your first queen's ghost, it should take joy / To see her in your arms' (5.1.79–81).

'Throw physics to the dogs': Hysteria or Possession?

Confronted with their respective wives' diseases, Leontes, Macbeth and Othello consider the possibilities of cure. Upon his wife's collapse, Leontes urges her attendants to 'apply to her / Some remedies for life' (3.2.150–1), still unaware that the latter actually depend on his good will. This explains Paulina's later insistence that 'no remedy but you will give me the office / To choose you a queen' (5.1.77–8), implying that her mistress's recovery rests on his disposition for marriage. Macbeth too entrusts his wife with a doctor whom he guides to examine her guilt-ridden conscience behind her hysterical agitation. His diagnosis mingles natural and unnatural causes as he addresses his wife's 'troubles of the brain' (5.3.43) in medicalised terms: 'Cans't thou minister to a mind diseased [. . .] / And with some sweet oblivious antidote / Cleanse the stuffed bosom of that perilous stuff / Which weighs upon the heart' (5.3.41–6). Macbeth's reading of his wife's hysteria corresponds to Timothy Bright's second type of melancholy which he distinguishes from natural melancholy and which he imputes to the affliction of sin. Except that, according to Bright, in this specific type of melancholy, the body and the senses remain sane and healthy, unlike in the first type, where they are obstructed by humours (Thomas Neely 2004: 15; Lund 2010: 57). Macbeth's approximate diagnosis echoes Leontes's own observation about Hermione: 'Her heart is but o'ercharged' (3.2.148). Both, therefore, link suffering to the immediate cause –

guilt over murder for Lady Macbeth and sorrow over a lost child for Hermione – but fail to perceive the very natural cause of the ailment: the disturbance of the womb and the ensuing imbalance of humours. Othello is the one who comes closest to the right diagnosis of his wife's disease as he intends his remedy to be a drastic antidote to lasciviousness. He believes that only death can be salutary to the sin of lechery and aims at easing Desdemona's unbridled womb and its unruly demands by means of a homeopathic remedy. Indeed, he bids her 'peace and be still' (5.2.49), in an endeavour to contain her uncontrollable 'erotic mobility' (Traub 1992: 39) of which her hot and moist hand is a telling sign. Rejoicing over her body, now a 'monumental alabaster' (5.2.5) irresponsive to his kisses, he admits to her complete purgation from corrupted humours and her eventual desexualisation: 'Cold, cold, my girl? / Even like thy chastity' (5.2.284–5). Her dead body becomes a mummy endowed with the therapeutic values of the *pharmakon* made from the embalmed corpse of virgins, 'an efficacious corpse drug contained and preserved in all its purity' (Noble 2004: 135). Hermione undergoes a similarly reparative transformation. Aged and wrinkled, she has indeed changed from the 'too hot' and all too fleshly 'goodly bulk' (2.1.20) into a 'warm' and 'sainted spirit' (5.1.57, 109), redeemed from the anxiety-breeding solicitations of the womb.

The search for a cure therefore informs the three male heroes' discourse which attests to the noticeable change of the age towards the medicalisation of the body. Each of them trusts the power of the *pharmakon*, be it 'poison' (4.1.190), Othello's initial 'swift means of death' (3.3.494), or 'remedies for life' (*The Winter's Tale*, 3.2.151) and 'oblivious antidote' (*Macbeth*, 5.3.44). They indeed attempt to decode and explore the hidden recesses of women's bodily parts at the cost of miscomprehension, couched in terms of wonder for Leontes who 'will not seek far' to understand his wife's miraculous reawakening (5.3.141), or even failure, as in the case of Macbeth who, confronted with the incompetence of the doctor, 'throw[s] physics to the dogs' for '[he]'ll none of it' (5.3.48).

In these early seventeenth-century plays, Shakespeare expresses the nascent trust in the power of science his era is witness to, and, in particular, the considerable influence Edward Jorden's treatise had on people's understanding of female psychophysiology. A fellow of the College of Physicians, Jorden centres his medical study on the highly topical case that captivated the attention not only of the citizens of London, but also of King James VI on the eve of his accession to the throne of England (Paul 1950: 107). The case is that of Mary Glover, a teenage girl allegedly bewitched by an old lady, Elizabeth Jackson, who caused her a formidable array of somatic torments daily, from spectacular writhing to comatose-like syncope (Macdonald 1991: xii; Almond 2004: 288–9; Paul 1950: 103–4; Thomas Neely 2004: 81–2; Levin 2002: 21–2; Sands 2004: 176–80). The debate over this case came to a head with the imprisonment of the presumed witch and resulted in a 'pamphlet war' (Paul 1950: 106). The king received two conflicting testimonies as to the validity of Mary Glover's bewitchment. Jorden's recently printed book, then consulted by the king, disrupted the supernatural interpretations of the girl's symptoms in favour of a scientific explanation of *hysterica passio* combined with melancholy, all due to an imbalance of humours (Jorden 1603: chap. 2; Macdonald 1991: vii–x; Thomas Neely 2004: 82). Grounding his study in the Hippocratic and Galenic theories, Jorden also advanced an innovative explanation to hysteria by extending its unique cause, i.e. sexual abstinence, to 'the perturbations of the minde', taking into account the fact that Glover was still a prepubescent (Jorden 1603: chap. 2; Macdonald 1991: xxviii). His medical thesis was supported by Samuel Harsnett in *A Declaration of Egregious Popish Impostures*, a religious diatribe against demonism and exorcism. Harsnett personally denounced Glover as a fraud to the king (Paul 1950: 107). Contrarily, the physician Stephen Bradwell and the Puritan minister John Swan wrote pamphlets supporting the stance that she was bewitched. The latter sent his *True and Briefe Report of Mary Glover's Vexation and her Deliverance* (1603) to the king with a dedicatory preface, relying on his support – being himself the author of *Dæmonologie* – for the discrediting of Jorden's thesis (Swan in Almond 2004: 287–8, 292–5). However, the king's enthusiasm about witchcraft and demonism had remarkably abated ever since he became acquainted with Jorden's rational explanations of women's pathological diseases (Gaskill 2006: 31, 106–7; Macdonald 1991: xlix). After having presided over witch trials in Scotland at the end of the sixteenth century, the new king of England now trusted the knowing look of the empirical observer Edward Jorden, whose treatise made him urge the release of the falsely accused witch from prison (Purkiss 1996: 201).

Nevertheless, the triumph of science over superstition was far from conclusive. For all his convincing explanations, Edward Jorden declined to cure the afflicted girl, which opened the floor for his opponents' criticism (Macdonald 1991: xvii; Almond 2004: 289; Sands 2004: 180–1). In one of the trials where both the victim and the tormentor were offered up as a spectacle, fuelling the contention of both camps, John Swan brushed aside Jorden's arguments in a way quite reminiscent of Macbeth's despairing renouncement of his faith in the doctor: 'I know they are learned and wise, but to say this is natural, and tell me neither the cause, nor the cure of it, I care not for your judgement. Give me a natural reason, and a natural remedy, or a rush for your physick' (Swan in Macdonald 1991: 29). Obviously, Jorden's 'natural reason' for Mary Glover's fits was adamantly turned down. Furthermore, six Puritan ministers, 'devil finders and devil puffers' in Samuel Harsnett's epithets (Almond 2004: 287), one-upped the physician by orchestrating a ritual of exorcism, with prayers and fasting, thanks to which, as Swan confidently details in his pamphlet, the girl was eventually restored to good health (Swan in Almond 2004: 291–330). Still, the Mary Glover case initiated an alternative understanding of women's ailments. Edward Jorden received further support from James I for the investigation of other cases of counterfeits, as in the case of Anne Gunter, in 1605, whose fraud he uncovered by medical means (Macdonald 1991: xlviii–xlix), just as later in the century, in 1634, William Harvey, the physician credited with the discovery of the circulation of the blood, won the approval of Charles I for similar missions (Aughterson 1998: 396–7). One of Harvey's most famous achievements was his clinical examination, with a large group of surgeons and midwives, of a woman allegedly possessed by the Devil, through which he invalidated the diagnosis of the marks on her body as a witch's teats (Gaskill 2006: 46–7). Hysteria and possession remained barely distinguished in people's minds, however. 'The Mother' became part of popular use, and Shakespeare's explicit naming of the disease and its symptoms in *King Lear* (1606) could not have been lost on its audience (2.4.54–5) (Almond 2004: 5; Gilman 1993: 100).

Likewise, his three other Jacobean plays resonate with the same topical debate between demonists and physicians and testify to the intersection of the spiritual and the material in the understanding of women's uterine diseases. Demonised and sacralised as hysterics, Hermione and Desdemona evolve in an opposite direction to Lady Macbeth, the hysterical witch (Levin 2002: 39), but the three embody the overlapping discourses of Satanism and hysteria. An unswerving believer in the power of a handkerchief and the 'magic in the web of it' (3.4.65), Othello is inclined to mix amateur knowledge in humoral theory and connoisseurship in matters of sorcery and exorcism. He examines Desdemona's hand after the fashion of both a physician, who reads the unusual emission of heat and moisture according to the Galenic pattern of the body's temperature (Paster 2004: 77–8, 85–6), and a witch-finder, who seeks the Devil's mark or *stigma diaboli*: 'Hot, hot and moist [. . .] here's a sweating devil here / That commonly rebels' (3.4.32, 36–7). The cure he prescribes to her is reminiscent of the demoniacs' in their treatment of Mary Glover: 'A sequester from liberty, fasting and prayer / Much castigation, exercise devout' (3.4.34–5). Moreover, his interpretation of the 'loves [Desdemona] bear[s] to [him]' follows the same blurring of distinctions between his initial discernment of 'appetite' (3.3.287), close to the scientific etiology of hysteria, and his subsequent moralising and religious reading of them as 'sins' for which Desdemona must die. The cure he orchestrates recalls the Catholic and Puritan sessions of confession as he asks Desdemona, 'Have you prayed tonight?' (5.2.26) and urges her, 'confess thee freely of thy sin' (5.2.57). Desdemona's response is just as ambivalent, soliciting first her Lord for supernatural deliverance (5.2.61, 87), then her lord, i.e. Othello, for natural remedy (5.2.129).

Hermione too is made to confess to her 'bolder vices' (3.2.54) but in a public trial, like witches. Even before the trial, her body – called a 'thing' (3.1.82) – is objectified and presented to the male gaze for scrutiny (2.1.64–5) (Rosenfield 2002: 102). Her anatomy is submitted to the rich taxonomy of Satanism as Leontes's humoral diagnosis of excessive heat falls short of developing (1.2.107). Instead, the cultural construction of women as 'devils' – playfully voiced by Hermione herself (1.2.81) – offers to Leontes a tautological articulation for his association of women with sin, lechery and witchcraft. He summons up the stereotypes with which the witchcraft treatises of the time are replete: Hermione readily falls into the category of the malevolent nurturing witch whose milk is infected (2.1.58–9) (Rosenfield 2002: 98), while Paulina, the 'mankind witch' (2.3.6), tantalisingly insinuates the long-established suspicion that the assistance and authority of midwives is associated with devilish practices (Harvey 1993: 81–2).[8] Unsaved in the trial by Apollo's oracle (3.2.125–6), Hermione is rescued sixteen years later by Paulina, the former witch turned priestess or thaumaturge, who couches the revivification of her mistress in religious terms: 'it is required / You do awake your faith' (5.3.94–5).[9] Paulina denies being assisted by 'wicked powers' (5.3.91) or being superstitious (5.3.43) but her ritualistic and miraculous enactment of Hermione's resuscitation, by commanding music and orchestrating the moves of the statue and its spectators, makes Leontes welcome the event as 'magic' (5.3.110). Hermione's medical remedy is absorbed by religious salvation, which

---

[8] On the equation of the witch and the bawd in *The Winter's Tale*, see Purkiss 2008: 67.

[9] On wondrous spectacle versus reasoned speech in *The Winter's Tale*, see Platt 1997: 165–8. On the Catholic discourse of the play, see Lim 2001.

points to the indeterminacy of both disease and cure, and possibly suggests that Hermione, like Lady Macbeth, 'more needs [. . .] the divine than the physician' (5.1.71). Indeed, the doctor in *Macbeth* proclaims his incompetence in curing Lady Macbeth, observing that her 'heart is sorely charged' (5.1.51) and that she is 'not so sick / As she is troubled with thick-coming fancies' (5.3.39–40). Much like early modern physicians such as Bright or Burton, his hybridised lexical expression mingles melancholy and sin, mental and spiritual disturbances. Rehearsing her and her husband's crimes through desultory monologues, Lady Macbeth chiefly confesses her feelings of guilt but also her sexual frustration, of which her sleepwalking is a symptom. The doctor does initiate a diagnosis of physiological disorder, observing through her 'slumber agitation' 'a great perturbation in *nature*' (5.1.9, my emphasis), but he quickly forgoes his attempt, believing that a religious cure would be more suitable. Furthermore, Lady Macbeth presents another feature that adds to the complexity of the nature of her disease. She is a witch attended by 'murd'ring ministers' (1.5.47) whom she urges to 'make thick [her] blood' and to 'take [her] milk for gall' (1.5.42, 47), a process that, whether fantastical or real, echoes the cultural constructions of the denaturalised body of the witch upon which the contemporary imagination thrived, like Leontes's, for example. As she evolves from witch to hysteric, Lady Macbeth retains some of the demonic characteristics, such as the 'damned spot' on her hand (5.1.33) which, beyond its being an imaginary stain of blood with the figurative significance of guilt, has a significance steeped in the cultural semantics of witchcraft. According to Ann Lecercle, 'to the Elizabethans it would no doubt have been the Devil's mark' (Lecercle 1991: 150), i.e. the spot by which the Devil seals the flesh of his worshippers and on which James VI writes at some length in his *Dæmonologie* (Lecercle 1991: 148). Lady Macbeth's sleepwalking scene is hence dominated by a three-fold discourse: of Bright's first and second types of melancholy – the one on the affliction of the soul and the one that unites Jorden's *hysterica passio* – as well as of witchcraft.

Conclusion: 'Is there any cause in nature?'

The intricate and hybridised representation of women's disease and cure in the three plays conveys the conflicting yet coexisting discourses on the topic between demonists and physicians in early modern England. Well informed about the current trends in the field of medical research, Shakespeare's works remain nonetheless impregnated with the cultural assumptions about witchcraft and devilry as well as with the religious frameworks of his time – both Catholic and Protestant. Hence, his complex approach to women's sexual afflictions, where scientific data is eclipsed by the diabolic, the superstitious, the sacred and the liturgical, testifies to the timid advancement of empirical investigation in his time, especially in the field of gynaecology, and is evidence of how science and medicine were not yet well-defined disciplines but still understood as natural and medical philosophies, with vague delineations from the supernatural in all its forms (Aughterson 1998: 346). In the Vesalian *theatricum anatomicorum* that his stage becomes in these and other plays,[10] his ailing characters offer themselves as case studies or as 'document[s] in madness' and in other afflictions (*Hamlet*, 4.5.178) to the investigating gaze of would-be anatomists who make use of the available discourses of the time to explain them, seamlessly weaving medical, cultural, moral and Christian terminologies. Perplexed by the animosity of his daughter towards him, King Lear calls for her anatomisation, conflating the emotional lexicon of the 'hard hearts' (3.6.36) and the scientific 'cause in nature that makes' them (3.6.34). Although left unanswered, his puzzled inquiry into the natural cause of hatred pins down the playwright's contemporary belief in the physical palpability of emotions and in the materiality of passion. Hence, next to lovesickness and melancholy, a wide range of affects have been mapped out in the field of the history of emotions where similar overlapping discourses mingling the scientific and the non-scientific have been examined. Sadness and grief have been the focus of recent studies in this field that disentangle the 'cause[s] in nature' from the other causes of these affects in Shakespeare's age and work.[11]

---

[10] On the anatomical examinations and dissections as public spectacles before and after Vesalius and on the Renaissance anatomy theatre, see Sawday1995: 64–5.

[11] See, for example, Sullivan 2013.

Chapter 4

'More, I prithee, more': Melancholy, Musical Appetite and Medical Discourse in Shakespeare's *Twelfth Night*

Pierre Iselin

In Shakespeare's *Twelfth Night*, the conventional discourse on abuses, music addiction and melancholy is identifiable from the first lines of the play. Medical doctors and divines of his time, following the Ancients, seem to agree that 'music is a roaring-meg against melancholy' (Burton 2016: II, 253), whereas they warn against its abuse. However, the treatment of the musical material onstage and the various discourses on music echo the voice of convention in a very oblique, not to say paradoxical, manner. First of all, it is the play with the greatest number of lines sung in Shakespeare's whole canon. It is also the only play that has a musical opening – not the traditional trumpet call, but more likely the music of a mixed consort. It is also a play whose musical epilogue confronts audiences and directors with the question of the limits of representation. So, more generally, one can wonder if the abundant stage music is in excess in Shakespeare's comedy. Is not the play as a whole stricken with 'melomania'? And in that case, for whom does the music sound?

In *Twelfth Night*, the melancholy lover's, the upstart's, the drunkard's and the profiteer's unhealthy appetites have, in turn, their musical expression, and musical expertise is shared by at least three of the characters: Feste, the professional jester and singer, Viola, the 'eunuch' who is supposed to sing 'high and low', and Sir Andrew Aguecheek, Olivia's would-be lover, who is advertised as playing the 'viol de gamboys'.

The closed world of the play – Orsino's court, Olivia's household and its kitchen – endlessly produces its own vicious circle of neurotic repetition, frustration, and posturing: Orsino, Olivia, Sir Toby,

Malvolio keep taking up poses as the melancholy lover, the mourner, the incorrigible roisterer, the spoilsport. In most of these dramatic constructions, real music is performed and becomes an object of commentary, if not of debate.

In the current discourses on abuses, those disorders and transgressions which music accompanies or causes, let alone its mercenary character, would certainly label it as excess. But paradoxically, the discourse *of* music, from the first to the last line of the play, may be seen as a commentary on the play and the limits of comedy, a reflection on listening and reception, creating an aural perspective in which the genre and mode of the song, the identity and status of the addressee, and the more or less ironical distance that separates them, interfere.

The object of the present chapter is to juxtapose the various discourses on these themes – medical, political, poetic, religious and otherwise – which circulate in early modern England, and to see how they intersect on Shakespeare's stage at a particular stage of the dramatist's career, when Robert Armin, the skilful actor-singer who must have played Feste, joined the Lord Chamberlain's Men after William Kempe's departure around 1600.

Music and Appetite: Classical and Contemporary Discourses

If madness and mental disorder are regularly expressed in the commonplace metaphor of musical dissonance,[1] the correlation of melancholy and the excessive indulgence in music, known as 'melomania', received much contemporary attention in several kinds of discourse – therapeutic, philosophical and political – and became a fashionable theatrical type, in adult as well as in children's companies, by the end of the sixteenth century (Austern 1992: 187–90). The musical treatment of melancholy, in particular, is discussed by Timothy Bright, a medical doctor and divine whose *Treatise of Melancholy* (1586) gives precise recommendations concerning both food and music:

> As pleasant pictures, and lively colours delight the melancholicke eye, and in their measure satisfie the heart, so not onely cheerfull musick in generalitie, but such of that kinde as most rejoyceth, is to be sounded in the melancholicke eare. (Bright 1969: 301)

---

[1] See *King Lear*, 4.7.15, and *Hamlet*, 3.1.156–7.

Such prescriptions are hardly followed by such stage-melancholics as Orsino or Jaques, who rather over-indulge in their consumption of songs and music. Immediately after the song 'Under the Greenwood Tree' and its burden have been sung by Amiens and the Foresters, 'Monsieur Melancholy' – so called by Orlando – insistently asks for an 'encore', expressing his desire to fuel his humour, a bulimic addiction expressed through the trope of appetite for music:

> Amiens: It will make you melancholy, Monsieur Jaques.
> Jaques: I thank it. More, I prithee, more. I can suck melancholy out of a song, as a weasel sucks eggs. More, I prithee, more! [. . .] I do desire you to sing. Come, more, another *stanzo* – call you 'em *stanzos*? (*As You Like It*, 2.5.9–15)

Likewise, Robert Burton, in his encyclopaedic *Anatomy of Melancholy* (1621), defines music as 'a sovereign remedy against despair':

> *Musica est mentis medicina mæstæ*, [music is] a roaring-meg against melancholy, to rear and revive the languishing soul; affecting not only the ears, but the very arteries, the vital and animal spirits, it erects the mind, and makes it nimble. Lemnius, *instit. cap.* 44. [. . .] [It] not only expels the greatest griefs, but it doth extenuate fears and furies, appeaseth cruelty, abateth heaviness, and to such as are watchful it causeth quiet rest; it takes away spleen and hatred [. . .] it cures all irksomness and heaviness of the soul [. . .] it is a sovereign remedy against despair and melancholy, and will drive away the devil himself. (Burton 2016: II, 253–4)

Because music – like food – is part of the treatment, its dosage must be controlled and its mode appropriate. In the case of love-melancholy, the remedy may easily turn into a poison, inducing or aggravating rather than curing the condition it is supposed to cure. Such is the ambivalence of musical therapy, theorised since antiquity in the notions of 'sympathy' and 'ethos' (mode).[2] Like his predecessor, Burton warns against the danger of music addiction, which can verge on a form of haunting obsession:

> As it is acceptable and conducing to most, so especially to a melancholy man. Provided always, his disease proceed not originally from it, that he be not some light *inamorato*, some idle fantastic, who capers in conceit all the day long, and thinks of nothing else, but how to make jigs,

---

[2] See Plato's *Republic*, III, 18.

sonnets, madrigals, in commendation of his mistress. In such cases music is most pernicious, as a spur to a free horse will make him run himself blind, or break his wind [. . .] for music enchants [. . .] it will make such melancholy persons mad, and the sound of those jigs and hornpipes will not be removed out of the ears a week after. (Burton 2016: II, 254)

Citing the same classical and biblical sources, but in an altogether polemical context, another contemporary discourse intersects the medical – the political status of music in the church and more generally in society. From Northbrooke to Prynne, radical Protestants have objected, not to music per se, but to what they termed its 'abuses'.[3] Precisely, it is the trope of food becoming a poison that Philip Stubbes, in his *Anatomy of Abuses* (1583), uses in his critique of music:

I say of Musick, as Plato, Aristotle, Galen, and many others have said of it; that it is very il for yung heds, for certaine kind of nice, smoothe swetness in alluring the auditorie to nicenes, effeminacie, pusillanimitie, & lothsomnes of life, so as it may be improperly compared to a sweet electuarie of honie, or rather to honie it selfe; for as honie and such like sweet things received into the stomack, dooth delight at the first, but afterward they make the stomack so quasie, nice and weake, that it is not able to admit meat of hard digesture. So sweet Musick, at the first delighteth the eares, but afterward corrupteth and depraveth the minde, making it weake, and quasie, and inclined to all licentiousness of lyfe whatsoever [. . .] being used in publique assemblies and private conventicles, as directories to filthy dauncing, thorow the sweet harmonie & smoothe melodies therof, it estraungeth the mind, stireth up filthie lust, womannisheth the minde, ravisheth the hart, enflameth concupiscence, and bringeth in uncleannes. (Stubbes 1882: vol. 1, 169–70)

Penetrating the soul through the ears – like food through the stomach – music arouses sexual desire and deprives man of his masculinity, making him 'womanish', as it rapes him ('ravisheth the hart'). But, according to Richard Leppert, 'what is most interesting is the sense that music is something which when consumed consumes in turn. Its danger lies in its being internalized: it infects us; it eats us out from within' (Leppert 1993: 87). The 'artificiall

---

[3] See my article 'The Apology of Music in England (1579–1605): The Paradox of a Polemical *Topos*' (Iselin 2014a).

shaking of the ayre' (Wright 1971: 168) thus has the power to penetrate the porous, permeable body of man and subdue him entirely,[4] all the more easily if the hearer suffers from humoral imbalance.

Stubbes likens the abuse of music to a medicinal substance mixed with honey ('electuarie'), or pure honey, which first delights, but rapidly causes disgust, surfeit and queasiness. In the form of an aural poisoning, absorption of 'sweet music' thus conduces to a contamination and pollution of the hearer, not only on an individual scale but on the scale of society at large – the worst kind of this degenerate art being the devilish conjunction of theatre and music, which is precisely the case in point in a play like *Twelfth Night*, where music is an object of debate on excess, correlated as it is with melancholy self-indulgence on the one hand, and inebriety on the other.

In one of his two chapters devoted to music, William Prynne, citing Augustine,[5] takes up the metaphor of digestion to denounce the perverse effect of depraved music:

> Wherefore [. . .] should we then walke delighted with vaine songs, that are profitable for nothing, being sweet only for a time, but bitter afterwards? For with such scurrilities of songs the intised mindes of men are effeminated [*animi humini illecti enervantur*], and fall away from vertue, flowing downe into filthinesse, and for those very filthinesses afterwardes feele paines, and vomit up that againe with great bitternesse which they have drunke downe with temporall pleasure, *etc.* [*cum magna amaritudine digerunt, quod cum temporali dulcedine biberunt*]. (Prynne 1633: 271)

Sweetness turns into bitterness, pleasure into pain, and music into an emetic. Even Burton, who defends music as a remedy against

---

[4] On the notion of penetration and soul loss (*alienatio mentis*), see Ficino's *In Timæus Commentarium*: 'Musical sound moves the body by the movement of the air; by purified air it excites the airy spirit, which is the bond of body and soul; by emotion it affects the senses and at the same time the soul; by meaning it affects the mind; finally by the very movement of its subtle air it penetrates strongly; by its temperament it flows smoothly; by its consonant quality it floods us with a wonderful pleasure; by its nature, both spiritual and natural, it at once seizes and claims as its own man in his entirety' (Ficino 1576: 1,453, my translation). For a burlesque treatment of the theory of penetration, see *Cymbeline*, 2.3.11–29.

[5] Sermo IX, 5, 'de decem chordis'.

melancholy, denounces the inebriating effects of its abuse, once again resorting to the same metaphor of digestion:

> Many men are melancholy by hearing music, but it is a pleasing melancholy that it causeth; and therefore to such as are discontent, in woe, fear, sorrow, or dejected, it is a most present remedy: it expells care, alters their grieved minds, and easeth in an instant. Otherwise, saith Plutarch, *Musica magis dementat quam vinum* [music maddens more than wine] [. . .] music makes some men mad as a tiger; like Alstolpho's horn in Ariosto; or Mercury's golden wand in Homer, that made some wake, others sleep, it hath divers effects: and Theophrastus right well prophesised, that diseases were either procured by music, or mitigated. (Burton 2016: II, 254)

The cycle of perverted appetite – stimulation, ingestion, satiety, disgust, queasiness – or that of inebriety constitutes a favourite trope of melomania, a pathological form of musical consumption. If 'the man that hath no music in himself' is predisposed to all sorts of vices, the one who consumes too much music loses manliness, courage, integrity, gender identity and sanity. 'Unrestrained appetite for and excessive indulgence in just about anything – food, drink, sex, sleep, poetry, music – could potentially emasculate man through lack of self-mastery' (Gibson 2009: 52). The discourse of excess, hinged on that of penetration, translates the melancholy disease and its corollary music addiction as a double alienation through the loss of self-mastery and that of masculinity. As Kirsten Gibson argues, 'the fear of excess was central to the masculine anxieties expressed in the discourse of melancholy disorder, and it also overshadowed anxieties about the attainment and maintenance of masculine social status more generally' (ibid.).

A third discursive layer must be considered at this stage. The preceding section has shown the characteristic ambivalence of music's effects as cause or result of melancholy, as remedy or poison, a complexity summarised in a paradox:

> Duke: though music oft hath such a charm
> To make bad good, and good provoke to harm. (*Measure for Measure*, 4.1.16–17)

That music operates mysteriously on man's mind and affections is a theory on which encomiasts and censors, physicians and philosophers agree. But how does music stir up these passions and move these affections? After reviewing the commonplace arguments through analogy (sympathy between the human soul and music), divine providence (the sounds that affect the ear produce a certain spiritual quality in the soul), physiology (the air in movement animates the vital and animal spirits) or mode (one musical ethos has one effect on the human psyche), Thomas Wright, in *The Passions of the Minde* (1604), offers a relativistic, quasi-phenomenological position, later echoed by Bacon:[6]

> in musicke, divers consorts stirre up in the heart, divers sorts of joyes, and divers sorts of sadness or paine: the which as men are affected, may be diversely applied: Let a good and a godly man heare musicke, and he will lift his heart to heaven: let a bad man heare the same, and hee will convert it to lust [. . .] True it is, that one kinde of musicke be more apt to one passion then another [. . .] Wherefore the naturall disposition of a man, his custome or exercise, his vertue or vice, for most part at these sounds diversificate passions: for I cannot imagine, that if a man never had heard a Trumpet or a Drum in his life, that he would be at the first hearing bee moved to warres. (Wright 1971: 168)

The effect of music therefore depends on the listener's culture, sensibility and morality as much as on its own intrinsic mode, Dorian, Lydian or otherwise. In his convincing and documented analysis of the play's musical events, David Lindley envisages the implications of such an individual reception of a musical stimulus (Lindley 2006: 199–218).[7]

This third context may account for Orsino's first word – 'if' – in the opening line of *Twelfth Night*: '*If* music be the food of love, play on' (my emphasis). The initial hypothesis – and there is much discursive virtue in 'if' – partly rests on the authority of Plato in establishing a genealogy of love, music being the creator of mutual love and sympathy (*Symposium* 187c). Then the association of love-melancholy and sad music plays to convention, if not fashion even. In a piece punningly entitled 'Semper Dolend, semper dolens', published in his *Lachrimae, or Seaven Teares* (1604) and dedicated to Queen Anne of Denmark, John Dowland poses as the melancholy

---

[6] See Bacon's *Sylva sylvarum; or, A Natural History in Ten Centuries* (1626), Century II, 'Of Music', in Bacon 1826, vol. 1, 291: 'But yet it has been noted, that though this variety of tune doth dispose the spirits to variety of passions, conform unto them, yet generally, music feedeth that disposition of the spirits which it findeth.'

[7] See also my article 'Music and Difference: Elizabethan Stage Music and its Reception' (Iselin 1995a).

musician – the malcontent he really was in life. The fashionable affec-
tation which had become a stage convention by the late 1590s was
an easily theatricalised mood in terms of miming action and sartorial
codes, but also in terms of musical repertoire, vocal or instrumental.

The fourth commonplace is that which correlates melancholy
with appetite(s). In his *De somno et vigilia*, Aristotle elucidates the
paradox of that lean bulimic, struck with insomnia, as the temperature of black bile hinders digestion:

> Now are the 'atrabilious' addicted to sleep, for in them the inward region
> is cooled so that the quantity of exhalation in their case is not great. For
> this reason they have large appetites, though spare and lean, for their
> bodily condition is as if they derived no benefit from what they eat. The
> dark bile, too, being itself naturally cold, cools also the nutrient tract,
> and the other parts wheresoever such secretion is potentially present.
> (Aristotle 1985: 3, 457a, 27–32)

The atrabilious' imperfect digestion is the cause of their excessive
appetite, since satiety, being deferred, leads to gluttony, then to surfeit
– the bear, inscribed in Orsino's name, is itself a well-known zootype
of inordinate appetite as well as an emblem of melancholy.

Diseased Desire in *Twelfth Night*

The inordinate appetite of the music-lover is the first that is voiced in
the form of a commentary on real consort music, obviously not the
kind of lively music prescribed by Bright:

> If music be the food of love, play on;
> Give me excess of it, that, surfeiting
> The appetite may sicken, and so die.
> That strain again, it had a dying fall.
> O, it came o'er my ear like the sweet wind
> That breathes upon a bank of violets,
> Stealing and giving odour. Enough, no more:
> 'Tis not so sweet now as it was before.
> O spirit of love, how quick and fresh art thou
> That, notwithstanding thy capacity
> Receiveth as the sea, nought enters there
> Of what validity and pitch soe'er,
> But falls into abatement and low price
> Even in a minute. So full of shapes is fancy
> That it alone is high fantastical. (1.1.1–15)

From Jaques to Cleopatra, via Orsino, the trope of music as food is consistently used in the context of melancholy:

> Give me some music; music, moody food
> Of us that trade in love. (*Antony and Cleopatra*, 2.5.1–2)

Like Jaques, Orsino asks for *more* music ('play *on*'), a cadence he wants to hear *again*, thus deferring the resolution of the final chord. Like the voracious ingurgitation that can only lead to disgust and queasiness ('surfeit'), which is a rather close approximation to the syndrome of *bulimia nervosa*, the melodic descent is interrupted and never ends ('Enough, no more'). Not surprisingly, the alternative suggested by Curio – 'hart' hunting – immediately causes the two motifs of melancholy and of inversion to resurface: the melancholy 'hart' – the very same on which Jaques moralises – becomes the emblem of sick desire in the figure of Actaeon pursued and devoured by his hounds:

> That instant was I turned into a hart,
> And my desires, like fell and cruel hounds,
> E'er since pursue me. (1.1.20–2)

Narcissistic desire, epitomised in the Ovidian paradox of 'My plentie makes me poore',[8] is thus doomed to morbid repetition, verging on self-destruction.

In Act 4, scene 2, the staging of musical repetition more explicitly reiterates the motif of death while Orsino and Viola/Cesario are comparing male and female passion, here again in terms of appetite ('appetite', 'surfeit', 2.4.92–102). Orsino asks for an 'old and antic' song that he and Cesario heard the night before, which 'did relieve [his] passion much, / More than light airs and recollected terms / Of these most brisk and giddypacèd times' (2.4.4–6), an opinion diametrically opposed to that of the physician:

> not onely cheerfull musicke in a generalitie, but such of that kind as most rejoyceth is to be sounded in the melancholicke eare: of which kinde for the most part is such as carieth an odde measure, and easie to be discerned, except that the melancholicke have skill in musicke, and require a deeper harmony. That contrarilie, which is solemne, and

---

[8] 'Inopem me copia fecit', Ovid, *Metamorphoses*, III, 587, Golding's translation (1567); see Golding 2002.

still: as *dumpes*, and fancies, and *sette musicke*, are hurtfull in this case, and serve rather for a disordered rage, and intemperate mirth, to reclaime within mediocritie, then to allow the spirites to stirre the bloud, and to attenuate the humours, which is (if the harmony be wisely applyed), effectuallie wrought by musicke. (Bright 1969: 347–8, my emphasis)

According to Orsino, the song has therefore worked as a remedy, not a poison. Instead of 'feeding' his passion, as in the opening scene, the song has salved it, he says. Musical therapy, however, seems to operate here in an odd – if not homeopathic – manner: the tune has been performed, possibly several times, on instruments before Feste arrives onstage to deliver the song once more, thus confirming the principle of repetition already observed, but though the musical mood is diametrically opposed to the 'light' music recommended by therapists, it seems to operate on the listener. One may only conjecture the original tune as closer to Dowland's 'Flow my Tears' (a tune later iterated in his *Lachrimae*) than to anything more lively. The text of the song itself (2.4.50–65) redundantly exhibits the metonymic attributes of death ('cypress', 'shroud', 'yew', 'flower', 'coffin', 'corpse', 'bones', 'sighs', 'grave'), reaching a form of textual surfeit through stuttering and repetition ('come away, come away', 'Fie away, fie away', 'Not a flower, not a flower', 'Not a friend, not a friend'). If one considers that the mood of grief is complacently established by the Fool's song and is directed at his principal listener, this form of sympathy verges on redundancy, or even solipsism, with the Duke's contemplation of his own corpse, and the elegiac celebration of his own death. But the fact is, as Lindley rightly argues, that if this 'incidental' music casts a light on the Duke's character, it is also addressed to Viola, 'for whom the possibility of dying without ever speaking her love is real [. . .] Orsino may have chosen the song, but it also speaks to and of Viola' (Lindley 2006: 207). While the Duke develops apparently contradictory arguments on his own constancy (2.4.19–20) and immediately concedes that men's affections are 'more giddy and infirm than women's are' (2.4.33), Viola indirectly confesses her own love for Orsino, resorting to the persona of a fantasised sister, and to the related figure of Echo:

Duke Orsino: How dost thou like this tune?
Viola: It gives a very echo to the seat
Where Love is throned. (2.4.20–2)

The song therefore may also be seen as prompting another image of love death than that of Orsino – that of Viola as a funerary monument in the guise of an imaginary alter ego:

> Duke Orsino: And what's her story?
> Viola: A blank, my lord. She never told her love,
> But let concealment, like a worm i' the bud,
> Feed on her damask cheek: she pined in thought,
> And with a green and yellow melancholy
> She sat like patience on a monument,
> Smiling at grief. (2.4.109–15)

Like the nymph, she is starved of her own voice, and has to resort to allegory to 'speak otherwise', but it is through the medium of the song that this veiled revelation occurs.

Therefore, if Feste's song exacerbates the passion of love and the humour of melancholy, and then contradicts the medical discourse, it is first because the logic of theatre is not that of medicine: the latter attempts to relieve pain and passions; the former exacerbates them and exhibits their extreme manifestations. It is also because the staging of the musical episode in this scene creates a complex perspective of reception. Music is played, then sung, and is finally commented upon. It certainly does not purge melancholy from its principal listener and addressee, but, even if the song may be perceived as a distanced, even ironical commentary on Orsino's excesses, it nevertheless allows Viola to improvise a new descant, a cryptic revelation which in time will provide the true remedy to melancholy, possibly audible in her name ('*Viola contra atrabilem*'),[9] or perceptible in the violet's symbolism of faithful love for a distant lover. Her charm is not that of song, but of the music of words. Olivia, 'addicted to a melancholy' (3.1.196) as she is, in no time converts her excessive mourning into an equally excessive passion of love:

> But, would you undertake another suit,
> I had rather hear you to solicit that
> Than music from the spheres. (3.1.106–8)

An erotic version of the game of musical chairs, with its off-beat rhythms and errors, *Twelfth Night* thus explores the mysterious paths of desire and insatiable appetite in the serious mode of love-melancholy.

---

[9] See Schleiner 1983: 137. Viola/violet was a commonplace remedy for melancholy at the time.

The (Under)World of Carnival and Lent

However, the world of Olivia's household shelters another cast with more earthy appetites, and feasting on an altogether more intoxicating music. The appetites are those of food, drink, easy money and mercenary gratifications of several sorts. Here also Feste – the licensed fool (1.5.59) – is allowed '*vox*', and offers what his audience expects of him, and what he is paid for. Contrary to the rowdy song 'of good life' one might expect, the first song that Sir Toby asks for is 'a love song, a love song' (2.3.36) which has been variously interpreted. Critics have seen it either as the pathetic spectacle of 'elderly lust, afraid of its own death' (Auden 1962: 521–2) or as an ironical commentary on the 'Duke's overwhelming but ineffectual mouthings, Viola's effective but necessarily misdirected charming, and of course Aguecheek's absolute incompetence as a suitor' (Hollander 1961: 232). If the identity of the addressee is uncertain, the song itself is a *carpe diem* meditation on the transitory condition of love, set to a rather melancholy tune, probably the melody used by Thomas Morley in *The First Booke of Consort Lessons* (1599), and William Byrd in a keyboard piece with variations found in the Fitzwilliam Virginal Book (c. 1619):[10]

> O mistress mine, where are you roaming?
> O, stay and hear; your true love's coming
> That can sing both high and low:
> Trip no further, pretty sweeting,
> Journeys end in lovers meeting,
> Every wise man's son doth know.
> [. . .]
> What is love? 'Tis not hereafter,
> Present mirth hath present laughter.
> What's to come is still unsure.
> In delay there lies no plenty,
> Then come kiss me, sweet and twenty.
> Youth's a stuff will not endure. (2.3.38–51)

The following commentaries concatenate puns and medical commonplaces ('mellifluous voice', 'contagious breath', 'dulcet in contagion'), fragments of catches – the musical emblem of inebriety on the stage – and ballads, which finally dislocate the ditty of the songs into

---

[10] See my 'Commentary on "O Mistress Mine"' in *Transmission and Transgression: Cultural Challenges in Early Modern England* (Iselin 2014b).

a dramatic exchange. Interrupting this 'rough music', the figure of Lent, disconcerted by musical excess, addresses the festive trio with accents of Puritan invective:

> Malvolio: My masters, are you not mad, or what are you? Have you no wit, manners nor honesty but to gabble like tinkers at this time of night? Do ye make an alehouse of my lady's house that ye squeak out your coziers' catches without any mitigation or remorse of voice? Is there no respect of place, persons nor time in you? (2.3.85–90)

Musical antipathy has ironical implications in the play, leading as it does to a more painful treatment, in and through song, as music becomes an instrument of mental torture at the hand of the triumphing Feste/Sir Topas later in the play (4.2). As Robin Headlam Wells puts it, 'with his natural suspicion of music and its powers of exciting the passions, the puritan is here symbolically trapped and caught by what he most "deprecates"' (Wells 1994: 219). Musical humiliation and provocation is then explicit when Malvolio is literally imprisoned, and Feste turns the knife in the wound in singing the ballad of 'Robin': 'My Lady is unkind [. . .] she loves another' (4.2.74–8). Here music takes part in, and is part of, the debate between the Fool and the Puritan, with the temporary victory of the stage-fool, and Puritan revenge looming large – with the actual closure of the theatres four decades later – in Malvolio's ominous conclusion: 'I'll be revenged on the whole pack of you!' (5.1.371).

Epilogue

The kitchen scene (2.3) thus has far-reaching implications, in the play and abroad, and cannot be summarised as the seasonal confrontation of allegorical figures (Carnival and Lent), or of the topical conflict of Merry Old England and Puritan England. Indeed, despite the complicity of the drunkards, even those who side with him libel Sir Toby's excesses:

> Feste: If Sir Toby would leave drinking, thou wert as witty a piece of Eve's flesh as any in Illyria. (1.5.27–8)

Later, Sir Toby must choose either to ban drinking or to be banned by his host (2.3), since his abuses make him a savage, 'fit for the mountains and the barbarous caves' (4.1.7). Even in the festive clan, voices denounce the abuse of excess, which is perceived no longer as the transgressive but temporary recreation allowed by the licence of Carnival, but as a permanent regression to savagery. In this world of sharp oppositions, the figure of licensed excess in the play is also that of the ubiquitous Feste, both actor and spectator, the virtuoso ventriloquist who musically exorcises the passions and humours of his sponsors or opponents, and projects on them, through either the expressive mood of the song or the commentary it gives birth to, an ironical distance. Stage music thus salves passions, like John Case's venomous toad – a pun on the Latin 'buffones'? – 'sucking the poison' from this music in the theatre:

> Comedy and tragedy are performed in the theatre not for citizens to imitate such effeminacy or savage barbarousness, but to learn how to avoid and flee those dangers, having watched other people's faults; then *if toads can suck the poison* from this music in the theatre which so vividly expresses those passions, it must be called a disease of humour and man, not a vice in the science or the art. (Case 1588: 61, my translation, my emphasis)

A figure of mediation between the two worlds of *Twelfth Night*, Feste is equally the mediator between the world of dramatic fiction and that of real life. His song ('When that I was and a little tiny boy', 5.1.382–401), a musical epilogue to the play, explores the limits of theatrical representation: is it an alteration of the traditional jig that concludes a play, or an integral part of the comedy? As with the initial musical episode, music here is loaded with the same semiotic and poetic ambivalence, and the worldly wisdom that it conveys is also the expression of a form of melancholy, a distant echo of the sophisticated consort music that opened the play. But this liminal piece is also an oblique commentary on the play at large and on the adversities of life, the rain and all.

Framed as it is by an initial and a final musical event, the play itself is punctuated, orchestrated and eroticised by music, whose complex effects work on the onstage, as well as on the paying audience. 'Free of even the scraps of traditional musical ideology that had been put to use in the plays preceding it' (Hollander 1961: 161), *Twelfth Night* inaugurates a form of aural perspective centred on reception – individual as well as collective reception – that heralds *The Tempest*.[11]

---

[11] On reception in *The Tempest*, see my article '"My Music for Nothing": Musical Negotiations in *The Tempest*' (Iselin 1995b).

Chapter 5

Saving Perfection from the Alchemists: Shakespeare's Use of Alchemy

Margaret Jones-Davies

One tender-hearted creature or other, Save Mercury and free him.
  Mercury Vindicated from the Alchemists at Court (Jonson 1963: vol. 7, 410, ll. 30–2)

When Francis Bacon describes the mind as an 'enchanted glass' that projects its inner harmony on to the celestial bodies, reducing their motions to 'perfect motions' (Bacon 2000a: 116), he is attacking one of the fallacies that impeded the progress of modern science. There is 'so differing a harmony between the spirit of man and the spirit of nature' that nature must cease to be the enchanted reflection of man's dream of perfection. The eventual discovery of flaws in the ordered universe paves the way to an empirical perception of the world. But before the word 'science' came to suggest an open approach to the study of nature, it could be applied to such varied fields as theology, as in St Thomas's *Summa*,[1] and the mysteries of alchemy, as Shakespeare's king in *All's Well that Ends Well* suggests in one of his rare uses of the word:

Plutus himself,
That knows the tinct and multiplying med'cine,
Hath not in nature's mystery more science
Than I have in this ring. (5.3.101–4)

---

[1] See Benoit 1991.

But as Bacon explains, ancient words must be conceived 'in a differing sense from that, that is received' (Bacon 2000a: 81), and so it is in the process of reinterpreting the word 'science' and severing it from the bonds of authority that its former ideal of perfection and wisdom is slowly transferred to what is only figurative and can no longer inform the study of nature.

An instance of the growing scepticism concerning the 'scientific' nature of alchemy can be found in Montaigne's *Apologie de Raimond Sebond*, where in already 'Baconian' words he shows how a man of importance tried to justify his quest for the philosopher's stone and the 'belle science' by referring to 'five or six' passages from the Bible (Montaigne 1999: 585).

Shakespeare shares the same scepticism regarding those arts that claimed the name of 'science', but he is not prepared to relinquish the dream of perfection which an old 'science' like alchemy embodied in its quest for a unity of contrary forces. Could perfection be saved from the old coherent vision of the world? Or was it doomed to disappear and leave the world bare and ruined as modern science developed?

Three centuries before, perfection was already an issue with William of Ockham (~1280–1349) when he resumed the position of the nominalists, who had paved the way to modern science by breaking the continuity between faith and reason, thereby freeing an empirical vision from the authority of revealed religion: against Thomas Aquinas's conception of a moral act as a perfecting of man's nature in concordance with divine perfection, Ockham questioned the possibility of man ever achieving anything approaching such an ideal, for in the same way as faith should be severed from reason, ethics should be severed from metaphysics.[2] But for all that, Ockham was not denying the relevance of metaphysical concepts, only putting them in their right place to stop them from intruding on the particular domain of an empirical approach to the world. Transcendental values were therefore not doomed to oblivion but should be expressed otherwise than literally. By theorising the opposition between the literal and the figurative (Ockham 1974: 220–1), Ockham was saving wonder-producing concepts from being dismissed as regressive superstitions and allowing them to maintain a redeeming function in a figurative sense.

---

[2] See Maurer 1962: 285–6: 'Ockham [. . .] severs the bond between metaphysics and ethics and bases morality not upon the perfection of human nature (whose reality he denies), nor upon the teleological relation between man and God, but upon man's obligation to follow the laws freely laid down for him by God.' Quoted by Marilyn McCord Adams in Spade 1999: 245.

By giving a substantial psychical reality different from an empirical reality to transcendental values, the figures of baroque poetry – a storehouse for metaphysical leftovers – can be seen as a distant consequence of the nominalists' far-reaching cleavage, and, in the same period, so can Shakespeare's apparently incoherent blend of scepticism and defence of transcendence. This subtle compromise, which is the stamp of Shakespeare's thought, is very different from the unconvincing idea that positivism can, by the mere wonder generated by scientific discoveries, satisfy the needs of the human soul, as Greenblatt suggests when he analyses the changing meaning of 'wonder' (Greenblatt 2011: 199)[3] – a debate that flourished in the nineteenth century with Matthew Arnold[4] and Coleridge's theory of imagination as a gateway to a transcendental vision.

When Hamlet experiences wonder in the contemplation of the perfection of man, inspired by Pico della Mirandola's first lines of his *Oration on the Dignity of Man* (1486), themselves a quotation from the hermetic book *Asclepius* (Ménard 1977: 120), he cannot help concluding: 'And yet to me what is this quintessence of dust?' (*Hamlet*, 2.2.297–8). As 'quintessence' is an alchemical term suggesting the fifth element – the perfecting of the opus when the four elements have been conjoined into a unity – the irony of linking it to 'dust', another positive alchemical term but used here in its usual sense connoting mortality[5] (Abraham 1998: 62–3, 75), brings out Shakespeare's ambiguous attitude towards alchemy. The equivocal use of these two words could suggest an ironical comment on the wonder first expressed by Hamlet and its hermetic flavour. Shakespeare could well be speaking of himself when he makes Lafew comment on those 'philosophical persons' who 'make modern and familiar, things supernatural and causeless' (*All's Well that Ends Well*, 2.3.2–3). Miracles are of times past indeed when the walking forest of Dunsinane is made out to be mere soldiers carrying branches, or when the

---

[3] 'It might seem at first that this comprehension would inevitably bring with it a sense of cold emptiness, as if the universe had been robbed of its magic. But being liberated from harmful illusions is not the same as disillusionment [. . .] It is knowing the way things are that awakens the deepest wonder' (Greenblatt 2011: 199).

[4] 'Mr. Darwin once owned to a friend that for his part he did not experience the necessity for two things which most men find so necessary to them – religion and poetry; science and the domestic affections, he thought, were enough. To a born naturalist, I can well understand that this should seem so [. . .] But then Darwins are extremely rare' (Arnold 1962: 979).

[5] Both terms suggest sublimation and perfection, 'dust' being 'the purified body of the stone', but 'dust' is used here disparagingly and suggests mortality.

moving statue of Hermione is but the real person pretending to be made of stone. And when he has the friar say 'let wonder seem familiar' after 'resurrecting' the 'dead' Hero (*Much Ado About Nothing*, 5.4.70), he could mean that the sources of wonder were in danger of running dry, and that there was no cause for wonder since Hero never died. But the phrase can also be understood as an intention of saving wonder from the prevailing scepticism of the times and allowing wonder to continue playing a part in the more secular world of the Renaissance. For it is the wonder occasioned by the unmasking of Hero that has allowed the moral conversion of Claudio, and made possible the perfecting of his love.

What was happening to the notion of perfection in a post-nominalist world? One can even wonder if it was safely tucked away for poetical use since, as Heinrich Wölfflin remarked, in the context of the baroque, the absolute is no longer perfection but the infinite.[6] Nowhere was it more plausible to find the remains of the concept than in the alchemical poetics that pervaded baroque poetry and art, for perfection was the founding concept of alchemy. The alchemy of the alchemists relied on the literal continuity between physics and metaphysics. This implies that for those who believed in its literal power, the imperfections of nature could be perfected: according to Paracelsus (1493–1541), for instance, the alchemist brings things to perfection. As Lyndy Abraham writes, 'during the seventeenth century, the ancient, philosophical debate of art versus nature was employed in expounding not only such subjects as education, gardening and cosmetics, but was also used in scientific discourse. Alchemy was viewed as an art that could perfect nature' (Abraham 1998: 10–11). But alchemy as a science was on the wane.

The Alchemical Ideal of Perfection Questioned

As Shakespeare plunders the treasures of alchemy in a post-nominalist fashion, he plays with the concept of perfection, pointing to its newly diagnosed frailty: 'When I consider everything that grows / Holds in perfection but a little moment' (Sonnet 15, ll. 1–2), defining himself as 'an unperfect actor on the stage' (Sonnet 23, l. 1) who, 'for fear of trust forget[s] to say / The perfect ceremony of love's rite' (Sonnet

---

[6] See Wölfflin 1952: 10: 'Instead of perfection and completeness, baroque art seeks movement, change, what is unlimited and colossal' (my translation). Quoted by Gérard de Cortanze (Cortanze 1987: 49, 80).

23, ll. 5–6). But he will not easily relinquish the necessity for the old
dream of perfection:

> shall I say mine eye saith true,
> And that your love taught it this alchemy,
> To make of monsters and things indigest
> Such cherubins as your sweet self resemble,
> Creating every bad a perfect best (Sonnet 114, ll. 3–7)

Satire against the alchemists was thriving (Abraham 1998: xx, bibliography; Healy 2011: 20–4). James was watchful and in his *Dæmonologie*
(1597) condemned 'Diuerse Christian princes' for indulging in those
'practiques' and thereby 'sinn[ing] heavily against their office in that
point' (James 1924: 24–5). In 1604 he reconducted the statute promul-
gated by Henry VIII in 1541 prohibiting witchcraft and alchemy. It is
the alchemists' ideal of perfection that comes under attack. In his collo-
quium *The Alchemist* (1524), Erasmus makes fun of old Balbinus who
deserves all praise and yet, as Philecous objects, no mortal can be said
to be absolutely perfect and what questions his perfection is precisely
the fact that his hobby is alchemy (Erasmus 2006: 61). Ben Jonson
debunks Mammon's admiration for the 'abstract riddles' of alchemy,
which only bring out the greed for gold that this passion reveals (*The
Alchemist*, 2.1.104), and in his 1616 masque, *Mercury Vindicated
from the Alchemists at Court*, Mercury himself comes forward to indict
'the sooty tribe' for their 'treason against nature' (Jonson 1963: vol.
7, 412, l. 108 and 410, l. 48).[7] It is the alchemists' ideal of perfection
that is ridiculed in the antimasque that follows his long accusation, as
'imperfect creatures, with helms of limbecks on their heads' enter, in
an attempt to torment Mercury. But Mercury appeals to Nature, who
appears personified 'in a glorious bower' while the Chorus prays to
her: 'Come forth, come forth, prove all the numbers then, / That make
perfection up' (ibid., 415, ll. 210–11).

---

[7] In this masque, Mercury is seen trying to escape from the tunnel of the furnace in
the 'Alchemist's work-house'. Vulcan and a Cyclope try to stop him. Mercury's
long speech is a call for help and an indictment of the art that betrays nature. In the
antimasques that follow, Vulcan enters with alchemists and 'imperfect creatures',
and finally vanishes, and the masque ends with the victory of Nature. Mercury
escapes from the alchemists and reclaims his power in the hermetical imagery of
the monarch. Jonson, like Shakespeare, keeps using alchemical imagery in a figurative sense: 'Jonson displaces the alchemical magus by a Mercurian monarch whose
art derives, we are to infer, from true magical communion with the gods' (Brooks-
Davies 1983: 91).

There is one example in Shakespeare's plays that seems to contradict an enlightened approach to the healing power of alchemy. In *All's Well that Ends Well*, the pure and virtuous Helena heals the ailing king of France by virtue of her art. The theme of the ailing king is an alchemical one since the stone is called 'king' (Abraham 1998: 110–12). The nature of her art, being advertised as superior to the Paracelsians' who declared him incurable, is unclear.[8] But an intertextual study brings out obvious parallels with Rabelais's satire of alchemy in the *Cinquième Livre* (chap. xxi), where Panurge and his cronies meet a woman named Entelechy, meaning Perfection in Aristotelian terms (*De Anima*, II, 1). Entelechy is the last stage of Alchemy, when the four elements are conjoined in the fifth essence or quintessence, the pure essence of the philosopher's stone, which can preserve from destruction; the elixir that can cure disease, rejuvenate, and even resurrect the dead (Abraham 1998: 75). In Rabelais, the rejuvenation is not due to the miraculous powers of the elixir but to a venereal disease of the type of pelada, which sloughs off the old skin. The sexual illusion is not lost on Shakespeare, who has Lafew qualify the cured king appearing with Helena as 'Lustig' (*All's Well that Ends Well*, 2.3.38) (Jones-Davies 1986: 75–6). So however pure and virtuous Helena may be, her art is made out to be nothing other than the dubious art of the alchemists, and the miraculous cure is but another instance of Shakespeare's secular reading of the art that James had condemned.

Needful Perfection

Satire against the literal achievements of the opus was one thing, but it was impossible to deny that the very sovereignty of the king had been built on metaphysical fictions that depended on modes of thought showing points in common with alchemical thinking. The function of the images relating to the figure of Hermes, the father of alchemy, had gone unquestioned for centuries before their re-emergence in the Quattrocento. Allusions to the Emerald Table, where the alchemical principles were expounded, were found for the first time in the thirteenth century in Albertus Magnus's

---

[8] Frances Yates considers Paracelsian alchemy as belonging not to the old superstitions but to the new Renaissance magic 'that expanded by alchemical and Paracelsist influence' (Yates 1975: 118). But Shakespeare makes sure by his Rabelaisian irony that a literal belief in Paracelsian miracles must be understood figuratively, and in that sense he shares Jonson's scepticism such as it is expounded in *The Alchemist*.

Mineralia.[9] Alchemical images, based on the notion of perfecting matter into gold, had been used for a long time in a political context, as they were the most adequate way of expressing the principles of sacramental kingship. When a king was crowned, his nature was supposed to change in the same way as the host was turned into the body of Christ during the process of transubstantiation (Strohm 1998: 61). As for the fiction of the two bodies of the king, the union of the private body with the public body, it was comparable to the alchemical notion of the union of contraries. Such images were needful to conceptualise the fiction of the two bodies of the king. Ernst Kantorowicz quotes the Italian lawyer Baldus, who in the fourteenth century described in his *Digestum* the *persona mixta* that conjoins contradictory qualities (spiritual and natural for instance) as a hermaphrodite (Kantorowicz 1957: 10), but did not point to the alchemical symbolism of this 'double thing' or *rebis*, which represents 'the perfect integration of male and female energies' that occurs in the production of the stone (Abraham 1998: 98). Another alchemical symbol is called upon in the construction of sovereignty. There is a need for an image of continuity to prove the statement '*dignitas non moritur*', and Kantorowicz comments on the example of the statute of Alexander III, *Quoniam abbas*, showing that when an abbot dies, his incumbency is transferred to his successor, and so it can be safely concluded that the abbot, like the king, never dies. The medieval jurists, he says, used the hermetic image of the phoenix, which was the perfect way to account for such impossibilities: the alchemical bird that never dies and is reborn from its ashes was an apt image.[10] The 'king', being equated with the 'stone', must go through a process of purification similar to the achievement of the opus, implying the alternation of opposite principles of dissolution and coagulation, amounting to a perfecting.

Shakespeare's histories rely on alchemical figures to found the legitimacy of the kings: the notion of a perfecting of kingship beyond the frailties, crimes or shortcomings of individual kings was too vital to leave aside. Bearing in mind that 'king' is another word for 'stone',

---

[9] Aromatico 1996: 137; Albertus Magnus, *Mineralia* (liber primus, tract I, caput III, 5): '*et hoc summum ingenium alchimicorum docet Hermes in* Secreto Secretissimorum suorum', that Alain de Libera interprets in his article 'Albert le Grand' as the Emerald Table (in Libera, Gauvard and Zink 2002: 27).

[10] See Kantorowicz 1957: 388 and following. Ernst Kantorowicz quotes from A.-J. Festugière, 'Le symbole du Phénix et le mysticisme hermétique', chap. VII, 3, p. 388 (see Abraham 1998: 152).

the example of *Richard II* is particularly striking. It brings out the necessary continuity between the king and the usurper, showing that the notion of kingship transcends the weakness of the one and the violence of the other. It would be tempting to analyse the play as the description of a break between two dynasties, and its usurpation like a revolution, but if the alchemical imagery is taken into account, there is in fact one king; when Bolingbroke becomes Henry IV he is just perfecting the image of the king that has been dissolved, for Richard is 'a king of snow / Standing before the sun of Bolingbroke' (*Richard II*, 4.1.250–1). This reading of the play questions a number of comments on the revolutionary nature of the usurpation. The castles along Richard's way of the cross stand for alembics (Abraham 1998: 31–2). Bolingbroke, like the philosopher's stone, is an infant, 'sorrow's dismal heir' (*Richard II*, 2.2.63) born of the queen. The image of the two buckets in the well typifies the notion of 'inversion' which is how the opus progresses (Jones-Davies 2005b). But perhaps one of the most striking alchemical images is that of the bird of Hermes (Abraham 1998: 25–6; Jung 1970: 228), which shows a continuity between the imagery in *Richard II* and that in *1 Henry IV*: for Richard is 'plume-plucked' (*Richard II*, 4.1.99), and it is after a long time that the young prince Harry, as the successor to Henry IV, is described by Sir Richard Vernon as 'feathered Mercury':
All plumed like ostriches that with the wind

> Baiting like eagles having lately bathed
> Glittering in golden coats like images,
> As full of spirit as the month of May,
> And gorgeous as the sun at midsummer,
> Wanton as youthful goats, wild as young bulls.
> I saw young Harry with his beaver on,
> His cuishes on his thighs, gallantly armed
> Rise from the ground like feathered Mercury. (*1 Henry IV*, 4.1.99–107)[11]

The contrary images of the 'plume-plucked king' and the 'feathered Mercury', applied to the corrupt king and to the son of his successor, are in fact stages in the process of perfecting the image of the king. It takes three kings to perfect the 'king'. Three kings embody the perfecting of the 'king' or stone, from Richard's corrupt reign, symbolised by the king eating up his feathers under the influence of a scorching sun (Bolingbroke being the sun), as is shown in the

---

[11] See also Jones-Davies 2005a: 110–11.

Ripley Scrowle (1588), to the glorious reign of the future Henry V, described as the bird of Hermes in all his glory when still a prince. The alchemical stories of the 'king', or stone, underlie the histories. George Ripley in his *Cantilena* tells of a childless king who must be dissolved in the *prima materia* symbolised by his mother's womb. Incest is an alchemical phase (Abraham 1998: 106–7; Jung 1970: 525). After phases of purification, the child born from this union becomes a great king and redeemer. In the later play *Pericles* (1608), similar alchemical 'ingredients' are concealed in the adventures of the 'king' (Abraham 1999: passim). Pericles sees himself as a man 'on whom perfections wait' (1.1.122). As he escapes from the court of the incestuous Antiochus he must be dissolved like a *rex marinus* in the mercurial waters of the sea before meeting with the alchemical queen Thaisa, separated from his daughter so as to be purified from all temptation of incest. The epic story of *Pericles* is once more the story of the perfecting of the 'king', or stone. Shakespeare's kings are alchemical kings, like Duncan, whose 'silver skin is lac'd with his golden blood' (*Macbeth*, 2.3.109).

The histories are held together by the alchemical figures that make up a pattern analogical to an opus, thereby redeeming the old necessary rituals from their medieval context by integrating them into the 'new' hermetic philosophy, allowing for a continuity between the Plantagenets and the Tudors and Stuarts. Throughout the histories, 'the glorious sun / Stays in his course and plays the alchemist, / Turning with splendour of his precious eye / the meagre cloddy earth to glittering gold' (*King John*, 3.1.3–6). There are those kings who grovel in the chaos of the stage of putrefaction – the dying King John sees himself as a 'clod / And module of confounded royalty' (*King John*, 5.7.57–8), as if the glorious sun had not played the alchemist and had not 'perfected' the image of the king. Richard III is another of those failing kings described as 'a base foul stone' (*Richard III*, 5.5.204). Even if the image suggests first and foremost jewellery, as it is said to be 'falsely set',[12] the alchemical innuendo is possible too, for the colour symbolism of alchemy is applied to the rebirth of the king in the person of Richmond, a homonym of Rougemont (the red mount), the castle near Exeter whose name terrified Richard since 'a bard of Ireland' had told him once that he would not live long after seeing 'Richmond' (*Richard III*, 4.2.107–9). For after passing from the stage of putrefaction or *nigredo*, the stone is purified during the stage of *albedo*, then it is red and gold in the phase of *sublimatio*.

---

[12] See the title of George Baker's alchemical treatise *The newe iewell of health* (1576).

This little anecdote has no other function than pointing to the alchemical imagery that underlies the theme of royal succession, for the Earl of Richmond will become the first Tudor, Henry VII.

The process of perfection at work both in alchemy and in the line of succession of kings is apparent in the description of alchemical or royal marriages: here again the conjunction of contraries, male and female, is the condition for the achievement of the opus which is otherwise 'unfinished' – an adjective applied to Richard, the Duke of Gloucester, the future Richard III (*Richard III*, 1.1.20). *In King John* the intended marriage between the Dauphin and the Lady Blanche is described in terms of an alchemical, mercurial marriage, as the image of the 'silver currents' stresses even more:

> He is the half part of a blessèd man,
> Left to be *finishèd* by such as she;
> And she a fair divided excellence,
> Whose fullness of perfection lies in him.
> O, two such silver currents, when they join,
> Do glorify the banks that bound them in. (*King John*, 2.1.438–43, my emphasis)

In Shakespeare's last history, Thomas Cranmer rejoices at the birth of Elizabeth, 'the maiden phoenix', and prophesies that she will, in spite of her virginity, guarantee her succession: 'Her ashes [will] new create another heir / As great in admiration as herself'; 'Those about her / From her shall read the perfect ways of honour'. In answer to this, Henry concludes, 'Thou speakest wonders' (*All Is True*, 5.4.40–2, 36–7, 55). The perfection of the future queen and her heir James puts an end to the tumultuous phases of the historical opus: as Douglas Brooks-Davies remarks, 'the hermetic myth lives as long as there was a historically effective belief in monarchical absolutism in Britain' (Brooks-Davies 1983: 8).

Hermetic Politics

If James chose not to alleviate John Dee's penurious last years, he did not hesitate to dedicate his justification of the Oath of Allegiance to the greatest patron of alchemists, Rudolph II (Patterson 2000: 89). For although the 'bible' of the alchemists, the Emerald Table, and the *Corpus hermeticum* (translated by Marsilio Ficino from Greek into Latin in 1471) were both supposed to have been written by Hermes

Trismegistus, it was fashionable to reject the first as superstition and to acknowledge the second as the basis for a new religion. With Paracelsus, the link between alchemy and the hermetic philosophy of the *Corpus hermeticum* is explicit (Yates 1964: 150). But the *Corpus hermeticum* had been the basis of Francesco Patrizi's *Nova de universis philosophia* (1591). It was possible to use the tenets of the work in a purely religious and political way, for without the magic (Yates 1964: 169) and in the context of the wars of religion, its defence of a world harmony could offer a common language for both Catholics and Protestants intent on bringing the wars to an end. For the king of Bohemia was as fascinated as James himself with the irenic project of a unified Europe that Giordano Bruno had inspired when he brought the tenets of the new philosophy of Francesco Patrizi to the countries he travelled through. He could recognise in hermetic wisdom the principle of a perfect harmony between the sublunar world and the heavenly spheres, between physics and metaphysics that he kept seeking in the stills of the alchemists. So whilst alchemy as a science was on the wane, its principles were still alive in the ideals of hermetic philosophy and could function as a metaphor of political peace, and help the transition from the closed world to the infinite world: it is striking that Copernicus associated his discovery with Hermes, just after the diagram showing the sun at the centre of the universe in his *De revolutionibus orbium coelestium* (Yates 1964: 154).[13] As Peter French says in his analysis of John Dee's *Monas hieroglyphica* (1564), 'the process of man's spiritual transformation is therefore the deepest subject of this work, rather than the mundane alchemical quest for gold' (French 1972: 77). So even if Dee's belief in the literal power of alchemy could not be questioned, there was space for a figurative interpretation of his work. Hermetic philosophy and the figures of alchemy pervaded all forms of culture. Ben Jonson, however ironical his stance against alchemy, decorated the fans of his masques with old Egyptian hieroglyphics[14] and in *Mercury Vindicated from the Alchemists at Court* he 'first establishe[d] James as the Mercurian monarch' (Brooks-Davies 1983: 90–1).

---

[13] Among the historians and critics who have studied the evolution from the old philosophy to modern science are the following: Burtt 1925, Koyré 1957, Rossi 1968, Webster 1975, and Debus and Walton 1997.

[14] To comment on the dance of the Tritons, he speaks of the inscriptions on their fans, one of them being 'a mute Hieroglyphick': 'Which manner of Symbole I rather chose, then [*sic*] Imprese, as well for strangenesse, as relishing of antiquitie, and more applying to that originall doctrine of sculpture, which the Aegyptians are said, first, to have brought from the Aethiopians' (Jonson 1963: vol. 7, 177).

Alchemical Patterns of a 'Perfecting' Process in the Romances

However different from one another, the romances seem to be obses-
sively following a structural pattern suggestive of the alchemical
phases of the opus that justifies their convergence under the same
heading: 'romances', 'late plays', 'tragicomedies', etc. (Mebane 1989:
181; Healy 2011: 197). All are concerned with the journey backward
or inversion which characterises the recovery of the *prima materia*,
or *mercurius*, made up of 'sulphur', or 'Sol', the male principle, and
'argent vive', or 'Luna', its female counterpart. Before achieving
perfection, the philosopher's stone must go through phases of dissolution, when the old form of matter is dissolved into the *prima materia*
(Abraham 1998: 153, 35), associated with the moon (making the
body a fluid or spirit), and coagulation, associated with the sun (the
reverse), which purify it. As the sea or the mountains symbolise
the state of *prima materia*, it is not surprising to find the characters
of the romances having to go through the process of dissolution in
such locations. The tempests in *Pericles*, *The Winter's Tale* and *The
Tempest* and the mountain in *Cymbeline* are the loci of the beginning of the redemptive process. Marina, 'born in a tempest', Perdita
wrecked on the 'shores' of Bohemia and Ferdinand on Prospero's
isle, or Imogen wandering into the mountains where her royal brothers were landlocked are the tragic moments of the phase of *nigredo*
or *putrefaction*, which initiates the stages of *purification* or *albedo*,
which will ultimately lead to the final happy endings of the *sublimatio* or *rubedo*, when, as in *The Tempest*, the tale will end 'set down /
With gold on lasting pillars' (5.1.211). The colours of alchemy, as
we have seen in the case of *Richard III*, are a clue to the presence of
the alchemical theme. It is no accident that the underlying fairy tale
of *Snow White and the Seven Dwarves* – an acknowledged source of
Cymbeline – has alchemical sources, with references to the colours
of alchemy (black, white, red and gold) and to the seven metals buried in the earth.[15] The three caskets in *The Merchant of Venice* are
made of lead, 'the *prima materia* of the Stone', silver, 'a symbol of
the *albedo*', and gold, 'the image of the sun buried in the earth', a
'perfect body' (Abraham 1998: 115, 183, 86; *The Merchant of Venice*,
3.2). Other symbols accompanying these phases – the alchemical
birds, the crow, the dove, the eagle and the phoenix – appear at one

---

[15] Robert Ambelain, *Dans l'ombre des cathédrales*, quoted by Ania Teillard (Teillard
1948: 213). See also Jones-Davies 2003: 208–13.

time or another in the plays. The symbolism of the black earth – 'the matter of the Stone during calcination or putrefaction' (Abraham 1998: 27) – that testifies to the stage of *putrefaction*, for instance, appears in the image of King John described as a clod of earth as we have seen, but also in Cloten's name in *Cymbeline*. Another alchemical sign of this stage is a nasty smell, characterising Cloten, who 'reek[s] as a sacrifice', and Caliban, who 'smells like a fish; a very ancient and fish-like smell' and who ends up in 'the filthy-man-tled pool [. . .] [that] / O'er-stunk their feet' (Abraham 1998: 184; *Cymbeline*, 1.2.2; *The Tempest*, 2.2.24–5 and 4.1.183–4).

Another sign of the alchemical process at work is the omnipres-ence in both *Pericles* and *The Winter's Tale* of the twin gods Apollo and Diana, the children of Jupiter and Latona ('the raw stuff of the philosopher's stone'; Abraham 1998: 114), who stand for the operations of coagulation (Sol) and dissolution (Luna) that repeti-tively punctuate the opus (Abraham 1998: 8, 54). In *Pericles*, the 'resurrection' of Thaisa takes place under the aegis of Apollo and his son Aesculapius (*Pericles*, scene 12, 108) in Ephesus, the famous location of Diana's temple. In *The Winter's Tale*, Apollo delivers his oracle, which is followed by the death of Mamillius, while Diana's lunar symbol, the bear, is responsible for the death of Antigonus (Jones-Davies 2010: 266–7).

The perfection attained in the happy endings of the romances is much influenced by the alchemical notion of perfection. The chemical wedding or *coniunctio* is the union of opposites: sulphur and mercury, sun and moon, body and spirit, male and female. The result of such a union is the philosopher's child or stone, 'an infant that has the power to conquer all disease and transform all things to perfection' (Abraham 1998: 149). This wedding takes place after phases of *sepa-ratio*. What is typical of the four marriages that bring the romances to an end (those of Marina, Imogen, Perdita and Miranda) is that they take place after moments of terrible strife and necessary stages of purification. All four heroines are the daughters of kings and dukes and therefore will wear the crown that the alchemical lovers always wear in the engravings showing the *coniunctio* or coitus. For in the romances, the chemical weddings are part of the process of the perfecting of the king or stone.

The perfection of the alchemical figure of the uroboros, a circle made by a serpent eating its own tail, brings out once more the para-dox implied in the achievement of perfection. For here the serpent both destroys and nourishes itself. An interesting version of this image can be found in an engraving of a manuscript by Leonhardt

Thurneissen zum Thurn, *Quinta essentia* (1574).[16] The serpent's head is a lion (sulphur, sun) and its tail an eagle (mercury, moon). Its body twines itself around the superposed sun and moon. This engraving helps to make sense of the otherwise obscure final vision of the soothsayer in *Cymbeline*, which tries to account for the paradoxically unequal peace between the Romans and the Britons, who, although they are victors, submit to Caesar. This conjunction of opposites breaks the usual cycle of revenge that followed victories in the histories. 'For the Roman eagle', says the soothsayer, 'From south to west on wing soaring aloft, / Lessen'd herself and in the beams o' the sun, / So vanish'd; which foreshadow'd our princely eagle, / Th'imperial Caesar, should again unite / His favour with the radiant Cymbeline, / Which shines here in the west' (*Cymbeline*, 5.6. 471–6). The heraldic symbol of the King of Britain was the lion and here Cymbeline is as 'radiant' as the sun. The imperial eagle 'lessens' itself to allow the submission of Britain to Rome so that a future union will be possible between the king and the emperor. The eagle vanishes into the beams of the kingly sun (or lion), which swallows it, as in the uroboros. The symbol helps the reader to visualise and so understand the victory of Britain in alchemical terms, an image of perfection which takes into account the unequal (imperfect) forces at work and thereby puts an end to the cycle of revenge.

Farewell to Art

When Prospero drowns his book he resorts to prayer in the Epilogue of *The Tempest*. Magic must be excluded from philosophy and theology since wonder must only be a reaction to the greatness of God. Augustine had rejected the magical parts of the hermetic books, and the attempt to Christianise the *Corpus hermeticum* supposed the dismissal of what was magical (St Augustine 1984: 330–7). The same conclusion as to the power of grace over everything is reached in the last words of Montaigne's *Apologie de Raimond Sebond*. Shakespeare comes to the same conclusion in the sonnets published in the same period as the last plays.

In her groundbreaking study of alchemy in Shakespeare's sonnets, Margaret Healy shows how most critics have neglected the alchemical lexicon since Alastair Fowler first discovered how the science of

---

[16] Zentralbibliothek, Zurich.

numbers gave form to the sequence (Healy 2011: 9–10, 94–6). Her reading of the last two sonnets (153 and 154) integrates obvious allusions to alchemy, dealing the last blow to years of fanciful interpretations. Taking up Peggy Muñoz Simonds's idea that Sonnet 153 is concerned with Eros while Sonnet 154 is concerned with Anteros, the cupid of sacred love (Muñoz Simonds 1986, quoted by Healy 2011: 128–9), Healy adds this allusion to Diana, 'the fairest votary' (154, l. 5), who doubles as the 'maid of Dian' (153, l. 2) but 'grows a bath' that gives a spiritual meaning to Marianus Scholasticus's anecdote.[17] She concludes that both types of desire (sacred and profane) are shown to be legitimate, and that Shakespeare's position on the subject is 'ambiguous' (Healy 2011: 128). I should like to contrast her reading to the idea that the two sonnets echo the conclusion of *The Tempest* – from where Venus is expelled – in the sense that priority is given to divine love rather than profane love. The first alchemical bath fails to 'cure' the lover, who is driven to 'drown his book' in his 'mistress' eye' (153, l. 14). But in Sonnet 154, the allusion to Diana is superseded by the biblical interpretation, which transcends the literal reading of the anecdote. Indeed, as Ronald Jaeger had shown, the last lines of 154 are a quote from the Song of Songs (8:7).[18]

The repeated rhymes 'prove'/'love' (153, ll. 5, 7; 154, ll. 13–14) suggest that the poet is out to return to the definition of a love that 'alters [not] when it alteration finds' in Sonnet 116, where the final couplet makes 'proved' rhyme with 'loved', and only divine love does not change. Shakespeare here again seems to be following Augustine, who rejected magic but praised numerology and counted certain numbers as perfect, such as number 7, which points to the perfection of divine love (St Augustine 1984: 465). Fowler's analysis based on texts from the Bible had shown and justified the importance of counting 153 sonnets rather than 154 (Fowler 1970: 90; Healy 2011: 94). I should like to add that a sequence of 154 sonnets does not lessen the perfection that the theory of numbers aims at. Indeed, the whole sequence of the sonnets is based on that perfect number 7. Macrobius's definition of 7 as a perfect number – perfect because it is the addition of an odd (perfect) number to an even (imperfect) one – gives us the key to the choice of 154 as the total number of sonnets: indeed, 154 is 7 x 22, with 7, the perfect number, redeeming the doubling of 2, the symbol of division. No wonder that the last two

---

[17] The source of the sonnet, a six-line epigram by Marianus Scholasticus, was printed in Florence in 1594. See Booth 1977: 533.

[18] *Notes and Queries*, 19.4, 125, quoted by Booth (1977: 538).

sonnets should be concerned with the elements and their conjoining, since in the same passage where Macrobius praises number 7 he shows how the respective qualities of the four elements can mingle into one another, as here, water with fire (Macrobius 1990: 100–1). Down to the odd-man-out sonnet, 126 (with only 12 lines), 18 sets of 7 sonnets can be found, and 18 is a multiple of 3. From 126 onwards 4 sets of 7 sonnets make up the remaining 28 (a multiple of 2, an even number adequately referring to the imperfect feminine, according to Pythagoras). Half of 154 is 77, which is yet another insistence on the importance of the perfect number, but Sonnet 77 is a doubling of the perfect number and adequately introduces the sequence of the rival lover. In a similar way, Sonnet 133 (a multiple of 7) announces 'a torment thrice threefold'.

As Falstaff said, 'They say there is divinity in odd numbers' (*The Merry Wives of Windsor*, 5.1.3). The 154 sonnets are a perfect locus for the borrowing of St Augustine's phenomenological account of the Trinitarian making-up of man in the *De Trinitate*, where the achievement of *cor unum*, or 'one', akin to the union of the three persons of the Trinity, is the aim, the *telos*. Augustine explores situations where the Trinitarian model is to be found in the human experience. Among the many examples that make up Books VIII to XV of the *De Trinitate*, at least three are taken up in the sonnets: the human family (father, mother and son) corresponds to the first 14 sonnets (a multiple of 7), then the triple alliance of the faculties of man (memory, intelligence, will/love) triggers in Sonnet 29 (lines 10–13, 'thy sweet love remembered') the importance of memory, which initiates the fifth set of 7 sonnets, making the trinity possible between the three Augustinian concepts *amans*, *quod amatur* and *amor* – which also structures Ramon Llull's (1235–1315) dialogue between the lover, the beloved and love in *The Book of the Lover and the Beloved*, included in *Libre de Evast e Blanquerna*. Llull speaks of the beloved's perfection when the Holy Trinity of his lover was revealed to him (Llull 1987: 92).

The religious influence of St Augustine during Shakespeare's last years coincided with the saint's popularity with both Catholics and Protestants in their attempt at putting an end to the wars of religion. When Prospero drowns his book of magic, Shakespeare also resorts to the perfection of numbers by using as many words as there are squares on the chess board, drawing a calligram of numerical perfection into the text (Jones-Davies 2015). And Ben Jonson too had concluded his rejection of the more sombre magic of the alchemists with the attempt to 'prove all the numbers then / That make perfection up' (Jonson 1963: vol. 7, 415, ll. 209–11) in his masque *Mercury Vindicated from the Alchemists at Court.*

Conclusion: Hermione's Wrinkles

The inclusion of alchemical images does not make Shakespeare less of a 'philosophical person', a defender of an age that was freeing humanity from the bondage of superstition and magic (Jones-Davies 2014: passim). But what had held the world together was not to be totally dismissed. Shakespeare's use of alchemical images shows that they can express realities which, having undergone the metamorphoses of secularisation, can still signify, and sometimes even create, new ways of thinking. Shakespeare does away with the magic of the 'imperfect speakers' (*Macbeth*, 1.3.68) but cannot reject the ideal of perfection that lingered in the Renaissance conception of man and found its best expression in numerology. At the end of *The Winter's Tale*, when Hermione is 'turned' into a 'dear stone' (5.3.24), Leontes finds that the elixir of alchemy had not miraculously saved her from the wrinkles that Time had drawn on her face. Alchemy doesn't work miracles in Shakespeare, but does create wonder.

Part III

Knowledge and (Re)Discoveries

Chapter 6

Of Mites and Motes: Shakespearean Readings of Epicurean Science

Jonathan Pollock

Judging from the derogatory use of the term 'Epicurean' in Shakespeare's plays, one might think that he shared Ben Jonson's disregard for this philosophical sect. However, closer analysis shows that the term is invariably employed by unsavoury or downright evil characters about those whom the playwright portrays in a more positive light. It is none other than Goneril who describes the effect Lear's train has on her house in such terms: 'Epicurism and lust / Makes it more like a tavern or a brothel / Than a graced palace' (*King Lear*, 1.4.222–4). It is Macbeth who berates his former allies thus: 'Then fly, false Thanes, / And mingle with the English epicures' (*Macbeth*, 5.3.7–8). And it is Pompey who, in his hope that Antony remain in Egypt, desires that 'Epicurean cooks / Sharpen with cloyless sauce his appetite' (*Antony and Cleopatra*, 2.1.24–5). Do these examples truly represent Shakespeare's understanding of Epicurean morals, or might he not have had a greater knowledge of ancient atomism than is usually recognised? If this is the case, then he was not the only Englishman of the period to take Epicurus seriously. In *The Anatomy of Melancholy* (1621), Robert Burton vindicates the Athenian philosopher, after centuries of calumny:

A quiet mind is that *voluptas*, or *summum bonum*, of Epicurus, *non dolere, curis vacare, animo tranquillo esse*, not to grieve, but to want cares and have a quiet soul, is the only pleasure of the world, as Seneca truly recites his opinion, not that of eating and drinking, which injurious Aristotle maliciously puts upon him, and for which he is still mistaken, *male audit et vapulat*, slandered without a cause, and lashed by all posterity. (Burton 2016: II, 248–9)

The very title-page and preliminary texts testify to Burton's (apparent) espousal of atomist philosophy, for he calls himself Democritus Junior. However, he goes to great lengths to explain why he 'usurped' the name,

> lest any man, by reason of it, should be deceived, expecting a pasquil, a satire, some ridiculous treatise (as I myself should have done), some prodigious tenent, or paradox of the earth's motion, of infinite worlds, *in infinito vacuo, ex fortuita atomorum collisione*, in an infinite waste, so caused by an accidental collision of motes in the sun, all which Democritus held, Epicurus and their master Leucippus of old maintained, and are lately revived by Copernicus, Brunus, and some others. (Burton 2016: 'Democritus Junior to the Reader', 22)

Burton's principal reference is not to atomist physics but to the apocryphal letters of Hippocrates, in which a melancholy Democritus is seen investigating the physiological causes of his humour, and whose incessant bouts of laughter, far from being a symptom of mental disease as Hippocrates first suspected, are a result of his extra-lucid appraisal of human folly.

Burton published his *Anatomy* after the playwright's death, but his historical sources were already available at the time Shakespeare was writing. For example, he quotes a third-century text by Athenaeus, the 'Dinner Experts' or *Deipnosophistae*, book 10, on Epicurus's particular way of treating those suffering from mental illness:

> For when a sad and sick patient was brought unto him to be cured, he laid him on a down bed, crowned him with a garland of sweet-smelling flowers, in a fair perfumed closet delicately set out, and, after a potion or two of good drink which he administered, he brought in a beautiful young wench that could play upon a lute, sing and dance, etc. (Burton 2016: II, 255)

The sick patient, bed, garland of sweet-smelling flowers, beautiful young wench and music are all elements of the reconciliation scene between King Lear and Cordelia at Dover (*King Lear*, 4.7). Is that to say that Shakespeare deliberately intended Lear's doctor to be of an Epicurean persuasion? If such is indeed the case, this therapeutic 'epicurism' is a far cry from Goneril's slanderous allegations.

Apart from passages in Seneca, Cicero and Athenaeus,[1] the two major sources concerning Epicurus and his doctrine during the

---

[1] Cicero broaches aspects of Epicurean thought in *De finibus, De fato, De natura deorum* and *Tusculanes;* Seneca in *Quaestiones naturales* and *De otio;* Athenaeus in Deipnosophistae.

Renaissance were a long Latin poem from the first century BC by a certain Titus Lucretius Carus, *De rerum natura libri sex*, and a compilation of Epicurus's letters and maxims made by a Greek doxographer from the third century AD, Diogenes Laertius, in the last book of his *De vitis dogmatis et apophthegmatis eorum qui in philosophia claruerunt.* Although neither work was published in England until long after Shakespeare's death,[2] Continental editions found their way into English libraries.

The *editio princeps* of *De rerum natura* appeared in Brescia around 1473. In 1511, Johannes Baptista Pius (Giovanni Battista Pio) published another edition in his native Bologna, followed by another in Paris in 1514. The text of Lucretius follows the Aldine edition of 1500 but Pius was the first to publish a commentary.[3] Denys Lambin (Dionysius Lambinus, 1520–72), professor of ancient Greek literature at the Collège Royal, produced a Parisian edition of *De rerum natura* in 1563.[4] This was the one possessed by Michel de Montaigne, who quotes (by my calculations) 438 lines of the poem in his *Essais*, duly translated and annotated by John Florio in 1603.[5] We know that Shakespeare consulted at least the English translation of *The Essays*, since Gonzalo's Commonwealth speech in *The Tempest* reproduces Florio's rendering of a passage from 'Des cannibals'.[6]

According to Stuart Gillespie, '[t]here is a serious dearth of recent editions and commentaries on Lucretius in British libraries and book inventories from this period – scarcely a single documented example of the work of Jansson, or Pareus, or Nardi is to be found'. Nevertheless, '[w]hile it may be true that Lucretius had had less impact on English writers than any other major Latin poet by 1650, he had been read and appreciated by an important minority' (Gillespie 2007: 242). This minority includes some of the most significant writers of the period: not only Edmund Spenser and Robert Burton, but also Sir Thomas Browne, George Chapman, John Donne, Thomas Hobbes, Ben Jonson, John Milton and Thomas Stanley. To

---

[2] John Pearson published the first English edition of the Greek text of Diogenes Laertius in 1664. Thomas Creech published a verse translation of *De rerum natura* in 1682, followed by an edition of the Latin text with a commentary in 1695.

[3] *In Carum Lucretium poetam commentarii a Ioanne Baptista Pio editi* (Paris, 1514). See Gordon 1962: 77–8. It is mentioned in Munro's edition (Lucretius 1866: I, 4–5) and in Bailey's edition (Lucretius 1947: I, 45).

[4] Dionysius Lambinus 1563.

[5] For more information, see Pollock 2015.

[6] See Frank Lestringant's chapter in the present volume.

this list should be added the Renaissance English 'virtuosi', wealthy amateurs committed to the advancement of natural philosophy. For example, Henry Percy, 9th Earl of Northumberland (1564–1632) and an exact contemporary of Shakespeare, patronised Thomas Harriot, Robert Hues, Walter Warner and Nathaniel Torperley. According to R. H. Kargon, the Northumberland circle was 'the only English school to combine Copernicanism with that complete rejection of Aristotelianism which accompanies acceptance of the atomic philosophy of Democritus, Epicurus and Lucretius' (Kargon 1966: 7). Likewise, Francis Bacon 'showed himself in the years 1605–12, in the *Cogitationes de natura rerum*, the *De principiis atque originibus*, and in the works added to his *Essays* in 1612, strongly inclined towards atomism, declaring it a "necessity plainly inevitable", and embraced too the concept of the void' (Gillespie 2007: 251). However, 'Bacon was to discard the atomic view of matter a little later: his main works favouring it [. . .] went unpublished for many years after his death in 1626' (ibid., 252).

It is less likely that Shakespeare knew Diogenes Laertius's doxography, although Latin translations would have been available. After the *editio princeps* of the Greek text in 1533, the first bilingual edition was published in Geneva in 1570 by the scholar Henri Estienne, with a fifteenth-century Latin translation by Traversari. 'A second bilingual translation came out in 1594, edited by Tommaso Aldobrandini and published by Cardinal Pietro Aldobrandini' in Rome (Tolomio 1993: 156). However, throughout the course of the sixteenth century there had been editions of Traversari's Latin translation published in Paris (1510, 1515), Basle (1524), Cologne (1535, 1542), Lyons (1541, 1546, 1551, 1559, 1561, 1566, 1592) and Louvain (1596). Tolomio explains that

> [t]wo other Latin editions appeared during the second half of the sixteenth century: one translated by Johannes Sambucus and published by Christopher Plantin in Antwerp in 1566, the other published in Paris in 1560 by Jérôme de Marnef and edited by a group of scholars [. . .] Lastly, there was a Latin edition published in Geneva in 1595 by Jacques Choüet. (Ibid., 157)

It is therefore not impossible, though (as we shall see) highly improbable, that Shakespeare had some knowledge of the only complete texts by Epicurus to have survived into the modern period, his letters to Herodotus (on physics), Pythocles (on meteorology) and Menoeceus (on ethics). As for Shakespeare's knowledge of *De rerum natura*, however, a considerable number of textual echoes suggest that not only did he draw upon Florio's translation of Montaigne, he also, like Ben Jonson, possessed his own personal copy of the poem.[7]

Atomus

In his 1546 translation of Polidore Vergil's *De inventoribus rerum*, Thomas Langley writes: 'Epicurus one of Democritus dysciples putteth two Causes Atomos or motes and Vacuitie or Emptiness; of these he saith the foure Elementes come' (1.2.4b). There are very few mentions of the Greek word *atomos* (Latin *atomus*) and its English equivalent in Shakespeare's plays. The most widely known is Mercutio's use of the word in his fanciful description of the fairies' midwife, Queen Mab, whose coach is 'Drawn with a team of little atomi' (*Romeo and Juliet*, 1.4.58). Given the immediate context, it would be more appropriate to speak of the atoms as animate mites, rather than as inanimate motes. In *As You Like It*, Celia tells Rosalind: 'It is as easy to count atomies as to resolve the propositions of a lover' (3.2.227–8). The technical name of the rhetorical figure employed here by Celia is *adynaton* ('impossible'), for it is strictly impossible to count atoms, not only because they are infinitesimal in size, but also because they are infinite in number. This is why the word is almost always used in the plural. An apparent exception is to be found in *2 Henry IV*, when Mistress Quickly inveighs against the First Beadle, calling him 'Thou *atomy*, thou!' (5.4.29, my emphasis). However, she is not suggesting that he is as minute and insignificant as a single solitary atom, but that he is as thin and emaciated as a skeleton, or *anatomy*. Being illiterate, she has mistaken the prefix *ana-* for the infinite article: *an*-atomy.

The paucity of references to atoms in Shakespeare's works suggests that he did not have first-hand knowledge of Epicurus's writings (those contained in Diogenes Laertius's doxography). However, Lucretius, in his attempt to acclimatise Epicurean doctrine to the Roman world, never uses the word *atomus* to translate *atomos*, nor even its strict Latin equivalent, *individuum* ('indivisible'). In order to explain what atoms are, he is accustomed to calling them the *matter* or *genital bodies* of things, and more often than not the *seeds* of

---

[7] A hypothesis that I began exploring in three previous papers; see Pollock 2009, 2013 and 2015.

things: 'quae nos materiem et genitalia corpora rebus / reddunda in ratione uocare et semina rerum' (*DRN*, 1.58–60).[8] When referring to 'the school of Leucippus and Democritus and Epicurus' in his essay 'Of Atheism', Francis Bacon will follow Lucretius's example and translate *atomoi* by 'seeds': he can scarcely believe 'that an army of infinite small portions or seeds unplaced should have produced this order and beauty without a divine marshal' (Bacon 2000b: 51). But Shakespeare's use of 'grains' in *Measure for Measure*, and 'germens' in *King Lear*, might also constitute allusions to atomist doctrine.

In Act 3, scene 1 of *Measure for Measure*, Vincentio, disguised as a friar, admonishes Claudio (who has been condemned to death by Angelo for indulging in premarital sex) in terms reminiscent not only of Genesis and the Anglican burial service, but also of Lucretius: 'Thou art not thyself, / For thou exist'st on many a thousand grains / That issue out of dust' (3.1.19–21). According to atomism, all things (i.e. *res*, reality) are material, the only immaterial substance being 'Vacuitie or Emptiness'. The soul, then, for as much as it exists, is not fundamentally different from the body: they are both made up of a chance congregation of atoms. Surprisingly, Vincentio makes no mention of the crucial distinction drawn by Marlowe's Doctor Faustus in his final hour: 'All beasts are happy, for when they die / Their souls are soon dissolved in elements, / But mine must live still to be plagued in hell. / [. . .] / O soul, be changed into little water drops / And fall into the ocean, ne'er be found!' (*Dr Faustus: The A-Text*, 5.2.110–19, in Bevington and Rasmussen 1993: 197). In fact the whole of Vincentio's speech is much closer in tone to Montaigne's essay 'That to Philosophise is to learn how to die' (1.20) than to orthodox Christian doctrine. Montaigne quotes Epicurus on death: 'Nor alive, nor dead, it doth concern you nothing. Alive because you are: Dead, because you are no more' (Florio 1999: 1.19). This is rendered by Lucretius as 'Nil igitur mors est ad nos neque pertinet hilum' (*DRN*, 3.830),[9] where the first and last words of the line, *Nil* and *hilum*, form the term *nihilum* ('nothing') which thereby encloses the word *mors* ('death'). Claudio is not to fear death, but to 'reason thus with life': 'Thy best of rest is sleep, / And that thou oft provok'st, yet grossly fear'st / Thy death, which is no more' (*Measure for*

---

[8] 'Which usually, in our explanations, we call matter, or the genital bodies of things, or the seeds of things', all translations from the Latin are mine unless otherwise stated.
[9] 'Therefore death is nothing, as far as we are concerned.'

Measure, 3.1.17–19). Lucretius says the same thing of sleep in a passage quoted by Montaigne in 'That to Philosophise is to learn how to die': 'Death is much less to us, we ought esteeme, / If lesse may be, than what doth nothing seeme' (Florio 1999: 1.19; *DRN*, 3.926–7).

Even when Vincentio appears to refer to the traditional Christian idea of the soul as the breath of life, he remains very close to the spirit of Lucretius. The latter defines the soul (*anima*) as a mixture of breath (*aura*), heat (*vapor*) and air (*aera*) which escapes from the body at death: 'Tenuis enim quaedam moribundos deserit aura / mixta uapore, uapor porro trahit aera secum' (*DRN*, 3.232–3).[10] Vincentio tells Claudio: 'A breath thou art, / Servile to all the skyey influences / That dost this habitation where thou keep'st / Hourly afflict' (*Measure for Measure*, 3.1.8–11). If this living breath were not material, it would not be subject to the 'skyey influences' that afflict the body. It is perhaps this selfsame material soul which Lear perceives on Cordelia's lips at the moment of her death.

Although surprising, and even shocking, in a friar, Vincentio's consolation speech is perfectly in keeping with the rest of his character. Shakespeare portrays the Duke as an Epicurean throughout the course of the play: not the lustful, whore-mongering Epicure that Lucio depicts him as (to his face, though unbeknown to Lucio!), but an adept of the pleasure-seeking philosophy vindicated by Montaigne and Burton. 'Pleasure [*hedonè*] is the beginning [*archè*] and the end [*telos*] of the good life', writes Epicurus to Menoeceus (Diogenes Laertius 1964: 2.391). Like the Greek philosopher, Vincentio distinguishes between natural and unnatural desires: 'Happy thou art not,' he informs Claudius, 'For what thou hast not, still thou striv'st to get, / And what thou hast, forget'st' (3.1.21–3). This is precisely what Nature berates Man for in a passage from Lucretius quoted *in extenso* by Montaigne in 'That to Philosophise is to learn how to die': 'semper aues quod abest, praesentia temnis' (*DRN*, 3.957).[11] The fourth book of *De rerum natura* has taught Vincentio that amorous passion is a source of displeasure, which is why he tells the Friar: 'Believe not that the dribbling dart of love / Can pierce a complete bosom' (1.3.2–3). Likewise, when he confides 'I have ever loved the life removed' (1.3.8), he is referring to

---

[10] 'Indeed the dying let out a breath mixed with warmth, warmth which also brings with it air.'

[11] 'You always want what you don't have and despise what you've got.'

Epicurus and 'that precept of his sect, HIDE THY LIFE, which forbideth men to meddle with public charges and negotiations' (Florio 1999: 2.16, 'Of Glory'). Unfortunately, as Duke of Vienna, he is not really at liberty to taste the joys of a life of retirement in a philosophers' Garden.

Vacuity

Lear's desperate appeal to the thunderbolts which threaten his bare head during the great storm on the heath may also contain a reference to Lucretius's *semina rerum*: 'Crack nature's moulds, all germens spill at once / That makes ingrateful man' (3.2.8–9). His retort to Cordelia in the first scene almost certainly does: 'Nothing can come of nothing'/'Nothing will come of nothing' (1.1.82 in Q1; 1.1.90 in F). 'Nothing can be made out of nothing' (1.4.131) is none other than the first principle of atomism: 'nullam rem e nihilo gigni divinitus umquam' (*DRN*, 1.150).[12] Or again, 'Nil igitur fieri de nilo posse' (1.205); 'de nilo quoniam fieri nil posse uidemus' (2.287);[13] 'At quoniam supra docui nil posse creari / de nihilo, neque quod genitum est ad nil revocari, / esse inmortali primordia corpore debent, / dissolve quo quaeque supremo tempore possint, / materies ut suppeditet rebus reparandis' (1.543–7).[14] Whereas the biblical God created the world *ex nihilo*, the ancient atomists deny that the gods had any hand in the creation and subsequent organisation of the universe. Everything is the result of fortuitous assemblages of atoms; but for these atoms to have chances of meeting and combining with one another, their existence must be eternal, their form immutable, their number infinite, and their movement incessant. Unlike the finite universe and the concentric cosmic spheres postulated by the Aristotelo-Thomist world picture, the atoms move about in 'an infinite deal of nothing', to borrow Bassanio's derogatory comment on Gratiano's conversation in *The Merchant of Venice* (1.1.114). Hamlet, too, leans away

---

[12] 'Nothing can ever be created by divine power out of nothing.'

[13] 'From nothing nothing can be born'; 'we see that nothing can be born from nothing.'

[14] 'But, since I have already shown that nothing can be created out of nothing nor any existing thing be summoned back to nothing, the atoms must be made of imperishable stuff into which everything can be resolved in the end, so that there may be a stock of matter for building the world anew' (trans. R. Latham).

from orthodox astronomical doctrine when he exclaims: 'O God, I could be bounded in a nutshell and count myself a king of infinite space, were it not that I have bad dreams' (2.2.252–3). According to medieval Christian theology, only the Creator is infinite, not His Creation.

Lear, then, finds himself in a world abandoned by the gods. 'Thou swear'st thy Gods in vain' (*King Lear*, 1.1.162), Kent tells him in no uncertain terms, and is banished for his plain-spokenness. Not that the gods do not exist; they exist well enough, but in the spaces between worlds, and have no regard for human-kind: 'Far above grief & dangers, those blest powers, / rich in their active goods, need none of ours' (*DRN*, 2.649–50, trans. Ben Jonson, in Greenblatt 2011: 305). 'Quae bene cognita si teneas', writes Lucretius, 'natura videtur / libero continuo, dominis privata super-bis, / ipsa sua per se sponte omnia dis agere expers' (2.1,090–2).[15] In a world bereft of Divine Law, only natural law pertains. Hence Edmund, alone: 'Thou, nature, art my goddess. To thy law / My services are bound' (*King Lear*, 1.2.1–2), a perverse echo of Lucre-tius's opening hymn to Venus: 'Goddesse, thou rul'st the nature of all things' (*DRN*, 1.21, in Florio 1999: 3.5, 'Upon some Verses of Virgil'). Lear is among those who ignore that 'everything has a limited power [*finita potestas*] and a fixed boundary [*alte terminus haerens*]; as a result they err, and their reason is blinded [*errantes caeca ratione*]' (*DRN*, 6.65–7). Regan makes it her duty to remind her father of his infirmity: 'O sir, you are old. / Nature in you stands on the very verge / Of his confine' (*King Lear*, 2.2.319–21). Lear's folly is perfect proof that, in Lucretius's words, 'When once the body by shrewd strength of yeares / Is shak't, and limmes drawne downe from strength that weares, / Wit halts, both tongue and mind / Doe daily doat, we find' (Florio 1999: 1.57, 'Of Age', and *DRN*, 3.451–4).

Driven hysterical by what he perceives as Goneril's lack of filial gratitude, Lear exclaims: 'O Regan! She hath tied / Sharp-tooth'd unkindness, like a vulture, here' (2.2.323–4). He thereby com-pares himself with Tityos, lacerated by vultures in the Underworld for having tried to rape Leto, the mother of Apollo and Artemis.

---

[15] 'If you bear in mind these truths, then Nature will appear to you free, utterly exempt from tyrannical sovereigns, and you will see that it is by her own means, of her own accord and by herself, without the concourse of the gods, that she does everything.'

Lucretius also mentions Tityos, but given that there is no after-life in atomist philosophy, the real 'Tityos nobis hic est, in amore iacentem / quem uolucres lacerant atque exest anxius angor' (*DRN*, 3.992–3).[16] The same is true of the other torments of hell, 'carcer et horribilis de saxo iactu' deorsum, / verbera, carnifices, robur, pix, lammina, taedae; [. . .] Hic Acherusia fit stultorum denique uita' (*DRN*, 3.1,015–23).[17] Indeed, Shakespeare seems to have written the tragedy of *King Lear* to bear out the veracity of these lines, transforming Lucretius's list of torments into so many stage properties. Upon recovering from his madness in a tent in Dover, Lear believes for a moment that, like Ixion, he is 'bound / Upon a wheel of fire' in hell; however, he is still very much of this world, and it is in this world that his tears 'Do scald like molten lead' (4.7.45–7).

The characters in *King Lear* are very far from having acquired that quietness of mind which the Epicureans called 'ataraxia'. Rather, most are 'minded like the weather, / Most unquietly' (3.1.2–3). This does not stop Lear from addressing Poor Tom, Edgar's mad beg-gar persona, as 'Noble philosopher' (3.4.168) and 'good Athenian' (3.4.176). Given Poor Tom's way of life and near nakedness, most commentators assume that Lear is equating him with the ancient Greek Cynical philosophers, Diogenes or Crassus. However, *De rerum natura* also celebrates a Greek mortal, '*Graius homo mortalis*', who dared oppose religious dogma; nothing stopped him, 'neque fama deum nec fulmina nec minitanti / murmure compressit caelum' (*DRN*, 1.66–8).[18] Although he is left unnamed, the reference is clearly to Epicurus. Lear takes the opportunity of asking the 'good Athenian': 'What is the cause of thunder?' (3.4.151). Is he correct in assigning to it supernatural causes; is it really 'the great Gods / That keep this dreadful pudder o'er our heads' (3.2.49–50)? Epicurus examines this question in his letter to Pythocles. In the last two books of *De rerum natura*, Lucretius also passes in review 'nubila, sol, imbres, nix, venti, fulmina, grando, / et rapidi fremitus et murmura magna minarum',[19]

---

[16] 'Tityos is here with us, lying prostrate in the throes of love, devoured by vultures and anguished anxiety.'

[17] 'Prison and the horrible fall from the Rock, whips, henchmen, the stocks, pitch, rods, red burning spits [. . .] In short, the life of Hell is here, in this life of fools.'

[18] 'Neither the reputation of the gods, nor thunderbolts, nor the threatening grumble from the skies.'

[19] 'Clouds, the sun, rain, snow, winds, lightning, hail, and sudden thunderclaps and great rumblings full of menace.'

only to conclude: 'O genus infelix humanum, talia divis / cum tribuit facta atque iras adiunxit acerbas!' (5.1,192–5).[20] Thunder and lightning, and all other meteorological phenomena, have purely natural causes; 'ignorantia causarum conferre deorum / cogit ad imperium res et concedere regnum' (6.54–5).[21]

Atomic Decomposition and Recomposition

When Gloucester happens upon Lear in the Dover countryside, he exclaims: 'O ruin'd piece of Nature! This great world / Shall so wear out to naught' (*King Lear*, 4.6.130–1). This is standard atomist wisdom. According to Lucretius, 'Mutat enim mundi naturam totius aetas, / ex alioque alius status excipere omnia debet, / nec manet ulla sui similis res: omnia migrant, / omnia commutat natura et vertere cogit' (*DRN*, 5.828–31).[22] During the storm, Lear himself strives to hasten the end of the world: 'You cataracts and hurricanoes, spout / Till you have drenched our steeples, drowned the cocks!' (*King Lear*, 3.2.2–3). And indeed, 'una dies dabit exitio, multosque per annos / sustentata ruet moles et machina mundi' (*DRN*, 5.95–6).[23] To support such a claim, Lucretius asks: 'Denique non lapides quoque vinci cernis ab aevo, / non altas turris ruere et putrescere saxa, / non delubra deum simulacraque fessa fatisci?' (*DRN*, 5.306–8).[24] Prospero will say much the same thing, after putting an end to the prenuptial masque in *The Tempest*: 'The cloud-capped towers, the gorgeous palaces, / The solemn temples, the great globe itself, / Yea, all which it inherit, shall dissolve; / And, like this insubstantial pageant faded, / Leave not a rack behind' (4.1.152–6) (*DRN*, 5.95–6).

---

[20] 'Oh unhappy human race, the day such effects were attributed to the gods, along with terrible bouts of wrath!'

[21] 'Ignorance of the true causes persuades [mortals] to confer upon the gods the empire over all things and to concede royalty.'

[22] 'Of th'universall world, age doth the nature change, / And all things from one state must to another range, / No one thing like it selfe remaines, all things doe passe, / Nature doth change, and drive to change, each thing that was' (Florio 1999: 2.12, 'An Apologie of Raymond Sebond').

[23] 'One day will suffice to bring about the ruin of the massy machine of the world, after years of being sustained in the void.'

[24] 'Do you not see stones themselves vanquished by time, high towers collapse and rocks crumble, the temples of the gods and their statues break apart from exhaustion?'

What is true of the macrocosm is also true of the microcosm. After his final defeat at the hands of Octavius, Antony confides in his servant Eros: 'Sometime we see a cloud that's dragonish, / A vapour sometime like a bear or lion, / A towered citadel, a pendent rock, / A forked mountain, or blue promontory / With trees upon't that nod unto the world / And mock our eyes with air. Thou hast seen these signs? / They are black vesper's pageants [. . .] That which is now a horse, even with a thought / The rack dislimns and makes it indistinct / As water is in water [. . .] My good knave Eros, now thy captain is / Even such a body' (*Antony and Cleopatra*, 4.15.2–13). In *De rerum natura*, clouds (*nubila*) are compared to the faces of giants (*Gigantum ora*, 4.138), a monstrous animal (*belua*, 4.142) and pendent rocks (*saxis pendentibus*, 6.195); 'Praeterea cum rarescunt quoque nubile ventis, / aut dissolvontur [. . .] / mittunt umorem pluvium stillantque' (6.513–15).[25] The clouds thereby lose all visible shape and become, in Antony's words, 'As water is in water' (*Antony and Cleopatra*, 4.15.12), 'umor ad umorem' (*DRN*, 2.1,114).

But just as clouds exemplify the processes which bring about the disintegration of all atomic structures, rainfall is associated by Lucretius with the emergence of new life: 'Quippe videre licet vivos existere vermes / stercore de taetro, putorem cum sibi nacta est / intempestivis ex imbribus umida tellus' (*DRN*, 2.871–3).[26] Accordingly, Antony swears 'By the fire / That quickens Nilus' slime' (*Antony and Cleopatra*, 1.3.69–70). Lepidus drunkenly expounds: 'Your serpent of Egypt is bred, now, of your mud by the operation of your sun' (2.7.26–7), and Cleopatra calls the asp 'the pretty worm of Nilus [. . .] / That kills and pains not' (5.2.242–3). The idea that putrefaction is the principle of life and nutrition stems from Aristotle and his theory of 'spontaneous generation'. If not in the *De rerum natura*, Shakespeare would have encountered it in the first book of Ovid's *Metamorphoses* (1.495–522); however, unlike Lucretius, Ovid makes no particular mention of worms or serpents. The passage in Lucretius continues: 'praeterea cunctas itidem res vertere sese. / Vertunt se fluvii, frondes, et pabula laeta / in pecudes,

---

[25] 'What is more, when, under the effect of winds, clouds fade away or are dissolved [. . .] they dispatch their water in the form of raindrops.'

[26] 'Do we not see live worms emerge from the repulsive slime, when the earth, drenched by excessive rains, has become rotten?'

vertunt pecudes in corpora nostra / naturam, et nostro de corpore saepe ferarum / augescunt vires et corpora pennipotentum' (*DRN*, 2.874–8).[27] Might not this rather sardonic example of the food-chain have inspired Hamlet's facetious remark to Claudius, 'We fat all creatures else to fat us, and we fat ourselves for maggots [. . .] A man may fish with the worm that have eat of a king, and eat of the fish that hath fed of that worm' (*Hamlet*, 4.3.22–8)? Everything is in a state of perpetual flux.

Conclusion: Shakespeare's Lucretius

Even if he had no access to writings by Epicurus, there is a strong likelihood that Shakespeare knew the *De rerum natura* by Lucretius, were it only via the numerous extracts contained in Montaigne's *Essays*, duly translated and annotated by John Florio. It has been our contention that the prevalence of weather images in Shakespeare's later plays is a result not only of his propensity for cloud-gazing but also of his interest in Lucretius's use of meteorological models in order to explain the creation and disintegration of material objects and living beings. Close readings of these plays reveal so many textual parallels with the original Latin that it is hard to believe in pure coincidence. Beside Shakespeare's Ovid and Shakespeare's Vergil, then, there is Shakespeare's Lucretius. But what are we to make of such an interest on the part of a (supposedly) Christian author? Epicurean science recognises only (atomic) matter and vacuity; it denies the reality of a spiritual 'substance' (God or an immortal soul). It would seem that Shakespeare uses Epicurean doctrine as a means of establishing dialectical oppositions: set against Lear's naive paganism or Cordelia's redemptive figure, atomism portrays a world without divine providence of any sort, prey to purely material forces; Antony's experience of his own dissolution contrasts with Cleopatra's vision of a cosmic Antony whose figure becomes immortalised in the stars; Prospero predicts the dissolution of the globe, while exerting his authority on spirits whose powers are denied by Epicurean rationalism. Shakespeare uses atomist physics and ethics

---

[27] 'Thus everything is turned into something else. Rivers and fronds and joyous meadows are turned into herds, herds are turned into our bodies, and our bodies, often, increase the vigour of wild beasts and the bodies of winged scavengers.'

in order to multiply perspectives and do justice to the complexity of human experience. Without rejecting Christian dogma, he places religious belief in a world where it is challenged by other systems of thought. Such metaphysical conflicts are rarely foregrounded (*King Lear* is an exception) but it is this dimension which contributes to making Shakespeare one of the most emblematic artists of the late European Renaisance.

Chapter 7

Shakespeare's Alhazen: *Love's Labour's Lost* and the History of Optics

Anne-Valérie Dulac

In a chapter from her *Study of* Love's Labour's Lost (1936) entitled 'The Earl of Northumberland and "Stella's" Sister' (Yates 1936: 137–51), Frances Yates defends the idea that Alhazen's optical theory lies behind many of the play's references to light, eyes and vision. She reads Shakespeare's early comedy as a dramatisation of the 9th Earl of Northumberland's attempt at quieting his sensual appetites and 'living in philosophy' (*Love's Labour's Lost*, 1.1.32)[1] through the creation of a 'little academe' of his own (1.1.13), devoted to the study of '[t]hings hid and barred [. . .] from common sense' (1.1.57). Yates's topical interpretation of the play as a satirical representation of Elizabethan scientific coteries was in line with previous 'exclusive focus on the play's topicalities' (Venet 2014: 275) by early twentieth-century commentators, which slowly grew out of critical fashion in the following years:

> All these hypotheses have long been dealt with as clever figments of what historical attitudes to literary texts could produce at times and were definitely dismissed as respectable oddities in more recent critical approaches to *Love's Labour's Lost* and mentioned as such in all recent editions of the play. (Ibid.)

---

[1] Unless otherwise indicated, all references to Shakespeare are to *The Oxford Shakespeare* (2005). I will therefore be using the names Biron and Mote, as appearing in this edition, to refer to the characters sometimes referred to as Berowne and Moth in other editions mentioned in the present chapter.

Although the view that Shakespeare's comedy was only intended as a mirror for mathematicians and philosophers somehow silences its wider complexity and poetic wealth, there is no denying that *Love's Labour's Lost* does reverberate some contemporary scientific theories and beliefs, which even recent editors never fail to mention – even if only *en passant*. For instance, commenting upon Biron's inaugural discussion of an optical conundrum – 'Light seeking light doth light of light beguile' (*Love's Labour's Lost*, 1.1.77) – William C. Carroll reminds readers that '[e]yes were believed to create the light by which they see' (2009: 65), thus concurring with Henry R. Woudhuysen's explanation ('[e]yes were thought to create (and project) the light by which they saw'; 1998: 117) and George R. Hibbard's suggestion ('it was believed in [Shakespeare's] day that [eyes] emitted "eyebeams" by means of which they saw'; 1990: 99). It therefore comes as a surprise that Yates's Alhazen hypothesis has failed to be examined thoroughly – even if only to be refuted – until very recently. I shall therefore first explore Frances Yates's assumption in greater detail, before considering the relevance – or lack thereof – of Alhazen's theory to the study of *Love's Labour's Lost*.

The Alhazen Hypothesis

Yates initially founds her assumption upon an autobiographical essay by the Earl of Northumberland dated circa 1604, which was published for the first time as an appendix to her own study under the title 'On Love'. In this text, Percy explains how, one day, as he was making himself 'giddy' by thinking about some unhappy love affair, he went rummaging in his personal library, in search of an alleviating read such as '*Tharcadia* or bookes of the like subjecte, whereby I might learne to vtter my lethargious passions with there sweete flimflams pleasinge orders' (Yates 1936: 207). But destiny, 'prepared to crosse [his] desires' (ibid.), had him stumble across an altogether different work by

> an owld Arabian called Alhazen, which wth some anger I angrylie removed, it flying open phapps by reason of a Stationers thred vncutt, yet superstitiouse in my religion that it was the spirit that directed me by hidden and vnconceavable meanes what was good for my purpose. (Yates 1936: 208–9)

Rather than indulging in the 'idle work'[2] of a poet, the Earl considers the incident as truly portentous and is then led – much against his own will, at first – to ruminate on how science may prove a surer way out of the 'circular maze' of 'fancies' (Yates 1936: 208). As a result, he passes from 'efforts to utter his "lethargious passions" in some kind of arcadian order' to a 'close application of his mind to Alhazen on light and on the theory of the rainbow' (Yates 1936: 147). This specific episode prompts Frances Yates to describe 'the theme of Northumberland's essay on the pursuit of learning [as] the theme of Shakespeare's play, reversed' (ibid.). Percy's decision to turn to 'philosophy' is indeed the result of disappointment in love, while Shakespeare's 'brave conquerors' (*Love's Labour's Lost*, 1.1.8) of knowledge are defeated by the Princess of France's embassy, announced only minutes after the four courtiers have pledged 'not to see ladies, study, fast, not sleep' (1.1.48).

Yates then carries the comparison further by implying that Biron's early lines on light may be a near-explicit reference to Northumberland's study of Alhazen's optics:

> Why, all delights are vain, but that most vain
> Which, with pain purchased, doth inherit pain:
> As painfully to pore upon a book
> To seek the light of truth, while truth the while
> Doth falsely blind the eyesight of his look.
> Light seeking light doth light of light beguile;
> So, ere you find where light in darkness lies,
> Your light grows dark by losing of your eyes. (1.1.72–9)

Yates takes the antanaclastic repetition of 'light' as evidence of the fact that Navarre and his men must have given pride of place to optics in their quest for knowledge, which she deems confirmed by Biron's subsequent development on astronomers (1.1.88–93), whose 'authority' in Renaissance Europe rested upon both mathematical and optical knowledge.

Yates's final point is that the influence of Thomas Harriot – a 'close student' of Alhazen's work and optics in general – is 'unmistakable in the Northumberland document' (Yates 1936: 147). Given that she

---

[2] These are words used by Philip Sidney to describe his own pastoral romance, the 'old' *Arcadia* (Sidney 1999: 3).

sees *Love's Labour's Lost* as not just alluding to Percy's essay but as a true mirror of the Earl's words ('[t]he play answers the document, the document answers the play, point for point'; ibid.), the conclusion she draws from these many reflections reads as a near-definitive statement: '[t]here can be little doubt, I think, that one of the books which the king and his court are studying at the commencement of the play is the *Opticae Thesaurus* of Hasan ibn Hasan or Alhazen' (ibid., 149). This became such a strong conviction of hers that Abdelhamid I. Sabra, who worked on his translation of Alhazen's *Optics* at the Warburg Institute, where he met Frances Yates, later fondly remembered the following anecdote:

> I have not forgotten the amused and kind smile on Otto Kurz's face when I expressed surprise at his interest in the *Optics*, nor shall I forget the day when Frances Yates walked into my room at the Institute with a copy of her 1936 *Study of* Love's Labour's Lost, in which she had pointed out a connection between the book by 'Hasan ibn Hasan or Alhacen' and the theme of Shakespeare's play. (Sabra 1989: vol. 1, xiv)

Strangely enough, however central Yates's reading of the play was, she never really discussed the plausibility of her hypothesis by looking at internal evidence, i.e. at the specific way the idea of 'light' appears in the play, to confront it with Alhazen's own theory.

Introducing Alhazen

Yates's very presentation of the work of Alhazen further confirms the fact that her deduction – which, I would like to show, was indeed a rather stimulating suggestion – is based upon very little inquiry into Alhazen's optics per se. When introducing the Arabic physicist to her reader, she writes:

> Hasan ibn Hasan, or Alhazen, was a celebrated Arabic mathematician of the eleventh century who made important advances in the study of optics. The book which the love-sick earl opened so carelessly was probably the Latin translation of his chief work, the *Opticae Thesaurus*, published at Basel in 1572. This book has an engraved frontispiece illustrating the subject. (Yates 1936: 141)

Against all odds, these lines are in fact the only passage in Yates's book about Alhazen proper. She never touches upon the very nature of the 'important advances in the study of optics' he is here said to have made. This surprisingly short presentation also features two rather inaccurate statements about Alhazen.

First and foremost, the title Frances Yates mentions, *Opticae thesaurus*, is in fact the title that was given by Friedrich Risner to a compendium that included an incomplete Latin translation of Alhazen's optical treatise, known in the West as the *De aspectibus* since the late twelfth or early thirteenth century, along with Witelo's *Perspectiva*. The first translation of Alhazen into Latin has been argued to be the work of probably two individuals, one of them having been identified (if hypothetically) as Gerard of Cremona. This is discernible through the fact that although the first half or so of the manuscript was relatively faithful to the Arabic text, the remaining chapters may be deemed 'a rather inept distillation of the Arabic original' (Smith 1987: xiv). The manuscript's translation errors and gaps reverberate in the commentaries of *De aspectibus* by Roger Bacon, John Pecham and Witelo. The 1572 edition that Yates refers to thus offers a version of Alhazen's theory (spelt for the first time with a 'z', whereas it had up to then been written 'Alhacen') that differs sensibly from the original, even though it was to become Europe's most common source of knowledge about Alhazen. As its title indicates, Witelo's famous text, which was appended to Alhazen's optics in the *Thesaurus*, belongs to what has retrospectively been termed the 'Perspectivist tradition', which gained popularity in the second half of the thirteenth century through the works of some key figures in the history of optics:

> In its narrowest sense, the 'Perspectivist tradition' comprises a select group of optical works that took their original inspiration from Alhazen's *Kital al-manazir*, which was written in Arabic during the early eleventh century and translated into Latin as *De Aspectibus* sometime in the early thirteenth. Aside from this seminal work, the tradition includes Roger Bacon's *Perspectiva* (1267?) and *De multiplicatione specierum* (early 1260s?), Witelo's *Perspectiva* (ca. 1275), and John Pecham's *Perspectiva communis* (ca. 1279). These treatises were enormously influential during the succeeding three centuries because all except Bacon's two works entered the university curriculum as texts for the teaching of mathematics and optics. (Smith 1987: 9)

The Perspectivist tradition thus owes its existence and influence to the common inspiration of its main authors from Alhazen's theory, to the point that historians of science usually consider their treatises as derivative works to the *De aspectibus*. It must therefore have seemed to be no coincidence to early modern specialists that Witelo's *Perspectiva* and Alhazen's *De aspectibus* were seen printed in one common thesaurus. This also shows how much Alhazen's thesis had been appropriated by the West, rather than just transmitted, to borrow the distinction made by Abdelhamid Sabra (Sabra 1994: xvi). Alhazen's work, although incompletely translated, was frequently, if not systematically, mentioned when dealing with optics in early modern Europe, if only in this Latinised form, different from the original.

The other slight inaccuracy appearing in Yates's introduction of Alhazen's *Optics* stems from the link she draws between the alleged contents of Alhazen's treatise and the frontispiece, or title-page, of the *Thesaurus*. Yates here asserts that the illustrated frontispiece adorning the compendium was intended to echo Alhazen's subject. In fact, the engraving incorporates elements from both texts in the thesaurus, some of which are actually absent from Alhazen's text and dealt with only by Witelo. Such is the case of the burning mirror, something that Alhazen did indeed study in other works but not in the *De aspectibus*. By 1572, the representation of a man concentrating the sun's rays on to a mirror facing ships sailing on the sea was a recurring cultural reference to the legend of Archimedes in Syracuse.[3] The previous editions of Witelo's *Perspectiva* (1535, 1551) had already been advertised as offering explanations on the design of burning mirrors, an optical marvel of sorts made popular by this episode, still featuring rather commonly on engraved illustrations of optical texts in the Renaissance. Although Frances Yates's summary is therefore partially incorrect, the influence of Alhazen's work and, as a result, its potential role in shaping Renaissance ideas on light, such as the ones voiced in *Love's Labour's Lost*, remains a plausible assumption. Examining such a possibility means probing into the exact nature of the 'optical advances' Alhazen made.

Alhazen relied on various sources, which he alternatively built upon or contradicted when elaborating his own theory. His main sources, although not all cited by names, have been positively identified as Euclid, Ptolemy, Aristotle and Galen. Alhazen refers to

---

[3] The legend has it that in 213 BC Archimedes successfully used burning mirrors to set the Roman fleet on fire during the Siege of Syracuse. The twelfth-century Byzantine authors John Zonaras and John Tzetzes were the first to comment upon this alleged optical miracle.

them as 'mathematicians', i.e. 'those who posit rays', which applies to Euclid and Ptolemy, 'natural philosophers' like Aristotle, and 'anatomists' such as Galen (Smith 1987: xxv). Alhazen's optics can thus be described alternately and simultaneously as mathematical, physical and physiological. What is most striking in this list is the encounter or confrontation of extramissionists ('who attributed vision to rays emanating from the observer's eye'; Lindberg 2002: 127) and intromissionists (who, following Aristotle, assigned 'the cause of vision to intromitted rays, which pass from visible object to observer's eye, where they stimulate the visual power'; ibid.). Alhazen's synthesis combined the two leading contradictory optical options of the time, thus granting him a long-lasting scientific posterity:

> Alhacen assigned the cause of vision [. . .] to intromitted rays [. . .] The rays efficacious in vision are those, he argued, that fall on the eye perpendicularly and enter without refraction, one from each point of the visible object. These, he demonstrated, form a cone of rays with the object as base and apex in the eye. At one stroke, Alhacen thereby joined the mathematical analysis of the extramissionists (associated with the visual cone) to the causal and physical concerns of Aristotle and the intromissionists. Set in its fully-developed form, within the anatomical and physiological framework of the Galenic tradition, Alhacen's theory achieved the unification he wanted. Championed by Roger Bacon, it dominated western thought until the seventeenth century. (Lindberg 2002: 127–8)

This unification of eclectic scientific traditions may be one of the reasons why Alhazen's theory did indeed dominate Western thought between the thirteenth and the seventeenth centuries. Alhazen's only potentially revolutionary claim could be made to fit into the evolution of Western optics from the Middle Ages to the early modern period mostly thanks to the Perspectivists' ability to accommodate his theory and weave it into their own still largely intromissionist model. Most of them mainly drew from the systematic features of his theories, like the 'idealized geometry of the eye', 'the selection of orthogonal rays by the lens' or the 'sensitive function of the visual spirit' (Smith 2001: cxiii). What Alhazen's approach showed was in fact that intromission did not stand in contradiction with extramission: despite their disagreement over the direction and type of radiation, the basic analytic device of extramissionist optics, the visual cone, was transmuted into a cone of light. Even in the extramissionist theory, vision is 'completed only when the passion of colouring is conveyed back through the visual flux to the eye' (Smith 2001: cxiv), so that for Ptolemy as much as for Alhazen, 'visual perception ultimately depends upon the transmission of illuminated colour from object to eye in the form of a cone' (ibid.), regardless of whether one posits a preceding visual ray emitted from the eye. The phrase 'eyebeam' accordingly applies to both theories and testifies to the coexistence of rival yet united views on light-beams and eyebeams, as may be best illustrated by Pecham's synthesis, as David Lindberg explains:

> [V]isual rays which are neither required for sight nor capable of producing sight by themselves, nevertheless play a contributory role, and that role is to moderate excessively bright lights so that they do not overwhelm the power of sight. Pecham thus yields to the authority of Aristotle, Alkindi, and Grosseteste on the existence of visual rays without seriously violating the teachings of Alhazen. (Lindberg 1970: 26)

Consequently, there was no contradiction in publishing Alhazen and Witelo together in a single volume in 1572, since Ibn al-Haytham had not only already been translated but also been adapted into European history. Although referring to the same person, the names Ibn al-Haytham and Alhazen may therefore be read as pointing to different takes on his optics.

Alhazen and *Love's Labour's Lost*

Interestingly enough, *Love's Labour's Lost* is the only play by Shakespeare where the word 'eyebeam' crops up as such (4.3.26). It is also true, as Yates points out, that 'eyes' and 'light' appear in just about every scene. This may be partly related to the fact that the play features many embedded sonnets and also parodies the clichés of the Petrarchan material. That sonnets and vision or optics are related is nothing new to literary historians and critics. Joel Fineman gives, for instance, the following description of this poetic format:

> The typical Renaissance sonnet embraces an idealised visuality which is reflexively reflective, on the model of the sun whose brightness is both agent and patient of both seer and seen, or of an *eidolon* whose

intromissive-extromissive visibility joins beholder to beheld, joining kind with kind, in an homogeneous conception of desire. (Fineman 1986: 20)

Fineman's typology usefully highlights the dual nature of visibility in the early modern sonnet, yet it does not give access to the full extent of the Shakespearean ambiguities at the heart of such dualism. This may be due to the fact that Shakespeare's own take on the sonnet was far from being 'typical'. My argument is that the 'homogene[ity] of desire' Fineman mentions as being characteristic of the sonnet is in fact questioned by its Shakespearean adaptation. Although mathematically equivalent, the two options (intromission and extramission) entail rather fundamental physical oppositions, a point that the very conception of light in both theories exemplifies quite well, with extramissionists positing light as a catalyst rather than as a direct object of sight – the latter which intromissionists, in the wake of Alhazen, would. Before his theory became known,[4] light could work as a catalyst between the seeing eye and the visible – as the two shared a common fire, according to the extramissionist theory – and perception rested upon a visual cone-like ray, acting as a sensitive projection from the eye to the world, much resembling a near-tactile beam engendering or bringing forth, quite literally, images. Indeed, as is made clear in Shakespeare's sonnets through the recurring use of the homophonic puns on 'sun/son', vision works as an analogy of procreation, of an encountering of same with same through which a shared light begets vision. With the visual beam emitted from the eye, the 'I' encounters the world outside the 'eye' and creates visions. The first sensorial encounter accordingly takes place outside the limits of one's body, if continuously, as appears in John Donne's poem 'The Ecstasy', which is very often quoted when discussing visual regimes in the Renaissance: 'Our eye-beams twisted, and did thread / Our eyes upon one double string' (Carey 1996: 113). The intertwining of beams celebrates the union of the two lovers who are, at the same time, and perhaps tellingly so, touching hands, in a possible evocation of the tactile dimension of eyebeams.

---

[4] I have already mentioned the many predecessors of Alhazen's theory and the way he built upon them in different ways to try to avoid drawing a Whiggish picture of the complex and non-linear history of optics. Although I am here focusing on Alhazen's work, I would like to make it clear that I am not considering his treatise as the result of individual genius and revolutionary insight, but rather as one among a constellation of texts and contexts which all contributed to the history of Renaissance optics.

The touch-like quality of extramission may also account for the much-eroticised value of the eyebeam metaphor so central to Petrarchan love poetry – hence, maybe, the resulting 'sun' of such sensory *and* sensual meeting. The analogy appears explicitly in *Love's Labour's Lost*, only to be humorously commented upon in a brief, yet revealing, exchange between Mote and Boyet, while the former is addressing the Queen of France and her ladies:

> Mote: Once to behold with your sun-beamèd eyes –
> With your sun-beamèd eyes –
> Boyet: They will not answer to that epithet.
> You were best call it 'daughter-beamèd' eyes. (5.2.168–71)

Boyet's punning, if intended as a joke, does in fact undermine the very extramissionist process in obliterating the fundamental analogy upon which it is grounded. Extramission can only work through comparison and analogy, with the light outside a father to his sons, engendered by similarly lit eyes. Gérard Simon, in *Archéologie de la vision*, makes it clear that the optics of 'those who posit[ed] rays' (Smith 1987: xxv) before Alhazen worked against a background implying a lineage between light, colour and sight. The lineage, or continuity, was such that seeing light was obviously the 'blind spot' in this theory, given that it would imply the capacity to see one's eye seeing, in an impossibly reflective act. This may be what actually fuels Biron's complex first lines in the play:

> Light seeking light doth light of light beguile;
> So, ere you find where light in darkness lies,
> Your light grows dark by losing of your eyes. (1.1.77–9)

Yet Boyet's punning, along with the rather evident irony directed against the would-be Petrarchs, somehow seems to deflate the artificial specularity of both rhetoric and vision, the 'heavenly rhetoric of the eye' (4.3.57).

Is there any alternative, any hint of intromission in the play that would not reflect homogeneity? In the intromission theory, the visible world affects the eye along lines of propagation coming from outside the eye, before they hit, quite literally, the crystalline (a movement which was sometimes deemed painful because of this sudden encounter between light – as agent – and the eye as receiving end of the process or *patient*, from the Latin for 'suffering'). As a result, intromission introduces discontinuity between the eye and the world, between the different moments of the perceptual process, between light and the body, thus making it possible, for example, for Alhazen to think of light as distinct from colour. The anatomists would also have supported such discontinuity. To borrow Sergei Lobanov-Rostovsky's enlightening formulation in 'Taming the Basilisk': 'during the sixteenth century, the practice of ocular anatomy made the eye visible to itself, intensifying the traditional conflict between the eye's material nature and its status as metaphor' (2010: 197). The 'vile jelly' of eyes could then be increasingly read about and seen, hence competing with the metaphorical lineage of the eye as agent of light, as father to son. Intromission may therefore be metaphorically interpreted as lineage gone awry, or obscure. And symptomatically enough, perhaps, in *Love's Labour's Lost*, the identity of the father of the only child in the play, Jaquenetta, remains a debated and unresolved point in the comedy – a blind spot, in other words.

Alhazen was also familiar with Galen and anatomy, and it is no wonder that his theory could incorporate the Galenic tradition. No longer simply understood as the sovereign sun, begetter of all sons, the eye was becoming this independent and material cell, only fantastical at times, receiving rays of light along mathematical lines. With light becoming a physical and mathematical object, the eye was conversely becoming more vulnerable and fleshy.

Is this the reason why Navarre's very first rules in the opening scene of the play state that the men should not *see* ladies or *eat* too much? Could it be that eating and seeing are paralleled because they are two forms of intromission? This brings to mind Maria Del Sapio Garbero's thought-provoking assumption in an essay devoted to *The Winter's Tale*:[5]

> Leonte's false knowledge, I argue, originates in what I would call a fatal short circuit between bowels and eye. Indeed, the alimentary trait and the eye are forcefully yoked together by Leontes when he says: 'I have drunk, and seen the spider' (2.1.45). He thus offers, through an intromissive, drinking metaphor, a marvellous synthesis of the two functions of seeing and corporeally eating up what he has himself created as a hallucinatory object: a spider. (2010: 136–7)

---

[5] I would like to express my gratitude to Maria Del Sapio Garbero for having so generously sent me her articles on the subject. They have been a truly valuable addition and incentive in my reflection on this topic.

This, in turn, may echo the female characters' decision in *Love's Labour's Lost* to turn their backs, rather than their eyes, to their suitors:

> Mote: 'A holy parcel of the fairest dames
> (*The ladies turn their backs to him.*)
> That ever turned their – backs – to mortal views.'
> Biron: 'Their eyes', villain, 'their eyes'!
> Mote: 'That ever turned their eyes to mortal views.' (5.2.159–62)

Shortly after refusing to grant the men their much-desired reciprocal gaze, the women will also, tellingly, expect them to fast:

> If for my love – as there is no such cause –
> You will do aught, this you shall do for me:
> Your oath I will not trust, but go with speed,
> To some forlorn and naked hermitage
> Remote from all the pleasures of the world.
> [. . .]
> If frosts and *fasts*, hard lodging, and thin weeds
> Nip not the gaudy blossoms of your love,
> But that it bear this trial and last love,
> Then, at the expiration of the year
> Come challenge me, challenge me by these deserts,
> And by this virgin palm now kissing thine,
> I will be thine [. . .] (5.2.784–99, my emphasis)

The lady's 'daughter-beamèd eyes' bestow no sun on the men's suit and condemn them to 'frosts and fasts' rather than the warm and fruitful season of love and procreation.

This may finally stand as one possible reason for Rosaline's many similarities with the dark lady of the sonnet: 'And therefore is she born to make black fair. / Her favour turns the fashion of the days' (4.3.259–60). In the sonnets, the dark lady is presented as a schismatic vision, undermining the sunny and homogeneous visions of the opening poems of the collections: 'Me from myself thy cruel eye hath taken' (Sonnet 133, l. 5). To quote Fineman again: 'the poet, with the lady, identifies himself not only with what is unlike himself, but with what is unlike itself' (1986: 22).[6] Contrary to the luminous

---

[6] I have discussed more thoroughly the optical metaphors in the sonnet and their potential intromissive dimension in my article on 'Shakespeare et l'optique arabe' (Dulac 2009).

exchanges and sun-basked reproductive visions of other sonnets,[7] the poems to the dark lady are grounded upon a painful and contentious visual process, whereby the continuity and fluidity of eyebeams are turned into a discontinuous trajectory between seer and seen.

Conclusion: Sources of Light

To conclude this chapter, I would like to turn to a footnote in the 2009 New Cambridge Shakespeare edition of *Love's Labour's Lost*, which offers a commentary upon Biron's most famous line on light: 'Light seeking light doth light of light beguile' (1.1.77). In order to unfold the many complex meanings and references behind such cryptic formulation, William C. Carroll writes that '[e]yes were believed to create the light by which they see, to search for light is therefore a kind of self-blinding' (2009: 65 n. 77). Yet this had ceased to be the only common optical belief for at least three centuries, over the course of which the supposedly 'dominant' theory had appropriated Alhazen's syncretic demonstration, if only in its Latinised form, substantially different from the original Arabic text. Although eyes were indeed believed to create the light by which they see, they were also thought to receive another kind of light, through intromission. Eyebeams could only give birth to visual perception when met by rays of physical light emitted from the world outside, exceeding the limits of the observer's bodily envelope. Such intromissive-extramissive theory of vision had resulted from the challenging history of the accommodation of Ibn al-Haytham's theory into what became known in Europe as Alhazen's *Optics*, after many a Perspectivist adaptation and reinterpretation.

This is the reason why Frances Yates's statement that '[t]here can be little doubt [. . .] that one of the books which the king and his court are studying [. . .] is the *Opticae Thesaurus* of Hasan ibn Hasan or Alhazen' (1936: 149) sounds rather too linear and evident today, in view of the recent developments and findings about the history of optics. The assumption that 'Hasan ibn Hasan or Alhazen' are only linguistically different names associated to one and the same theory rests upon the idea that the Latinised version of Alhazen is an exact translation of the medieval Arabic urtext, which it is not. Centuries of Perspectivist tradition had already incorporated the

---

[7] See, for example, Sonnets 3, 7 and 18.

original text into an eclectic synthesis that integrated varying positions on the subject of eyebeams and rays of light. This does not mean that Yates's intuition and conviction are irrelevant since Alhazen, if only in his European version, was widely read and woven within the Latin tradition, which allowed his views to become fully compatible with jarring notions, making his *De aspectibus* a dominant source of Western optics and, as such, a source as plausible as any when looking into Shakespeare's eyes.

Chapter 8

Shakespeare's Montaigne: Maps and Books in *The Tempest*

Frank Lestringant
Translated by Sophie Chiari

Performed at the court of James I on the evening of All Saints' Day, 1 November 1611,[1] *The Tempest* opens with a shipwreck at a time when scientific and public interest in cartography was on the rise. The opening scene of Shakespeare's romance begins with this maritime crossing, or at least its sudden end, and it goes on to present the multiple facets of a topos that was fashionable at the time, that of the pastoral. Exiled on an island that he now rules, Duke Prospero reigns over a lush world, a vision of fertility and natural abundance. After the shipwreck, life and the world start anew, and the quest for knowledge that had cost Prospero his dukedom now becomes an intellectual and spiritual pursuit embraced both by the derelict magus and by his old and not-so-naive counterpart, Gonzalo. So, as a play probing the meanings and the limits of knowledge, *The Tempest* intertwines two major intellectual issues of the time, namely the necessity of domesticating remote spaces to map and conquer the earth, and the importance of humanism as a (non-exclusive) source of scientific thought.

Geographical Prologue

The play is replete with geographical allusions that often tend to be contradictory. Milan, Naples, Algiers and Tunis, mentioned in turn, are not just cities but exotic worlds that give the spectator the

---

[1] For further details on the play's first recorded performance, see Laroque 2011: 173.

occasion to discover and dream of far-off places. Yet contemporary audiences could also occasionally find themselves in familiar London, especially when, having just discovered the 'strange beast' (2.2.30) Caliban, the Neapolitan jester Trinculo refers to the English display of natives from the New World: 'Were I in England now, as once I was, and had but this fish painted, not a holiday-fool there but would give a piece of silver' (2.2.27–9).

However, in spite of a few specific geographical hints, the play as a whole originates from a fantasised cartography, since its plot unravels on a mysterious island in the Mediterranean (a sea with no clear interior limit), a mythical Mediterranean which encapsulates the unknown, the wondrous, but firstly, due to the tempest and the shipwreck at the beginning, the deepest of Hells. Incidentally, Caliban may well be one of its infernal creatures since he bears hellish features as well as a devilish name: 'Caliban' is a near-anagram of 'Cannibal', the designation for that exotic savage whose geographical presence in the sixteenth century was imperceptibly shifting from the West Indies to South America.

In Renaissance-era atlases, a gaping, tempestuous Hell was pictured in the world's southern territories. In Petrus Apianus's *Cosmographia*, a highly successful work published no fewer than sixty times in various languages before 1600, a map of the world, drawn by Gemma Frisius, represents the winds and their effects: they could indeed bring fertility or illness, prosperity or plague.[2] The winds themselves are shown in three distinct ways: at the top of the map as the heads of bearded old men, in the top left- and right-hand corners as tousled and curly-haired children blowing out small suns, and at the bottom as skulls with dishevelled hair and sharp teeth, some spitting flowers and some smaller skulls. It would require but a minimal transformation to see these hot, evil and deadly winds replaced by fire-breathing devils.

---

[2] Finished as early as 1540 and inserted as a reduced and simplified version in the *Cosmographia* from 1544 onwards, Gemma Frisius's *Charta cosmographica* was republished for more than forty years. It was notably framed by an illustration representing the twin figures of Jupiter riding an eagle and of the Emperor Charles V with a sword and a breastplate decorated with the two-headed imperial eagle. In the 1551 Latin edition, this cosmographical map is placed between sheets Ai v° and Aij r°; in that of 1584, it is situated between pages 72 and 82. For further details on Gemma Frisius's map (inspired by a map of the world drawn by Petrus Apianus in 1520), see Hallyn 2008: 45–53, 66–7.

The Renaissance was a time when people were fascinated by the foreign and the unknown. No wonder, then, that in *The Tempest* Caliban the native regards himself as the real master of an island whose characteristics merge the Brazil of the cannibals – as depicted by Montaigne twenty years or so after the disappearance of Nicolas Durand de Villegagnon's 'Antarctic France' (1555–60)[3] – with the Edenic features of Arcadia. The New World depicted by Shakespeare is in fact a nascent world both worse and better than before, adorned in its most beautiful finery, and yet already riddled with troubles and conspiracies. From this perspective, *The Tempest* would, to some extent, present a new Genesis followed by a new Fall, were Prospero not vigilant enough. For, in Shakespeare's play, humanity is given a new beginning after a brief and partial flooding, a submersion from which redemption emerges or, at least, a new life much resembling the previous one, with a few exceptions. At the end of *The Tempest*, preparing for their departure from the island of this new Genesis, Prospero does not find this new humanity to be any better than that which he had to flee twelve years earlier. He eventually bets on a new beginning and takes a ship bound for the mainland, to Italy. He prepares himself to recoup the lost dukedom of Milan, of which he still is the legitimate owner, with his unrepentant brother Antonio by his side.

The Tempest does not simply issue from a somewhat arbitrary topography: it is also an indirect testimony of the scientific discoveries of the time and can therefore be seen as a reflection of early modern cartography, in which imaginary peoples and geographical references effortlessly coincide. For instance, the Austral continent seemed lavish in Shakespeare's time. An interesting correlation can be seen with the lush vegetation and hideous creatures (not unlike the revolting slave Caliban) prominent in Guillaume Le Testu's 1555 *Cosmographie universelle* (*Universal Cosmography*) of the region, which depicts monstrous races haunting distant lands. Incidentally, Le Testu, a Protestant cartographer and explorer, was a friend and travel companion of Sir Francis Drake. It was likely that Shakespeare had never heard of him, but he set his play in a geographical landscape very similar to the one described by Le Testu.

---

[3] See Florio 1999: 1.30: 'I have had long time dwelling with me a man, who for the space of ten or twelve yeares had dwelt in that other world, which in our age was lately discovered in those parts where Villegaignon first landed, and surnamed Antartike France.' All references to Florio are taken from the online edition.

So, one may here introduce the various maps of Le Testu's *Universal Cosmography* as possible precursors to the Shakespearean geography. Indeed, like most of his French and English contemporaries, the playwright had a rather restricted vision of these new horizons. Yet, basic as it is, this strange and fragmented vision turns out to be fascinating since it focuses on the New World before its birth and its individuation, i.e. before its complex separation from the Old. Take, for instance, Le Testu's map of Mexico with its golden, two-headed eagle in flight, standing between the two columns of Hercules, whose scroll reads '*Plus oultre*' ('still beyond') (Le Testu 2012: f.LIIIv°), or his three maps of the mythical Terra Australis, in which white unicorns are seen bouncing from the undergrowth (Le Testu 2012: f.XXXVv°, f.XXXVIIIv°, f.XXXIXv°).

Do not such oddities call to mind Antonio the usurper, vanquished by Prospero's charms, or Alonso's brother Sebastian, marvelling at the sight of these odd spirits of 'several strange shapes' (3.3.19, s.d.,) who bring forth a banquet and, through graceful dancing, amiably invite them to partake? 'A living drollery!' (3.3.21) he exclaims in amazement. Having just seen a life-sized puppet show, he adds, spellbound:

> Now I will believe
> That there are unicorns; that in Arabia
> There is one tree, the phoenix' throne, one phoenix
> At this hour reigning there. (3.3.21–4)

Yet, for all the geographical richness of these much-referenced lines, I propose to move beyond Shakespeare's fantastical overtones to evoke rhetorical, philosophical and political issues – for geography and legends, in *The Tempest*, serve to conceal a daunting political fiction.

Indeed, Ariel and his spirits fly over a territory where human passions convey the turmoil of air and sea. Once rescued from the shipwreck, the refugees on the island are quick to revive their old criminal habits. It is in this particular context that Montaigne is quoted by Gonzalo, Alonso's 'honest old counsellor', as Shakespeare calls him.[4] Far from giving up his fantastical geography, the playwright now adorns it with rhetoric. Drawn from the depths of collective memory, the vestiges of classical antiquity, and imbued with a new setting, Shakespeare's geography becomes surprisingly evocative, its historical richness revealing a surprisingly relevant undercurrent, renewed for the times with a touch of irony.

---

[4] See Shakespeare's *The Tempest* ('The Persons of the Play').

Guillaume Le *Testu's Terra* Australis (f.XXXVv°), © BnF.

A 'negative formula'[5]

In Act 2, scene 1 of the play, a tirade by Gonzalo, a learned and benevolent gentleman, is directly borrowed from Montaigne's 'Of Cannibals' (1580), an essay that Shakespeare came across in John Florio's 1603 translation. By revisiting this almost literal borrowing, already firmly established as such by scholars, my intention is two-fold. On the one hand, with the benefit of hindsight, I would like to propose a new reading of Montaigne's essay through Shakespeare's dramatic use of this passage. On the other, I intend to reassess Gonzalo's role through the prism of Montaigne's text in order to rehabilitate the old counsellor and his scholarly culture.

The negative formula, an inescapable trope of primitivist discourse encountered in Montaigne as well as in Shakespeare, is what actually structures Gonzalo's monologue in *The Tempest*.[6] In effect, the rhetorical device used by the old lord is the same as the negative formula found in Montaigne's well-known essay on cannibals, which reads as follows:

> C'est une nation, dirai-je à Platon, en laquelle il n'y a aucune espèce de trafic; nulle connaissance de lettres; nulle science de nombres; nul nom de magistrat, ni de supériorité politique; nul usage de service, de richesse, ou de pauvreté; nuls contrats; nulles successions; nuls partages; nulles occupations, qu'oisives; nul respect de parenté, que commun; nuls vêtements; nulle agriculture; nul métal; nul usage de vin ou de blé. Les paroles mêmes, qui signifient le mensonge, la trahison, la dissimulation, l'avarice, l'envie, la détraction [au sens de 'calomnie'], le pardon, inouïes. Combien trouverait-il la république qu'il a imaginée, éloignée de cette perfection: *viri a diis recentes*?[7]

Florio's 1603 translation rather faithfully reproduces Montaigne's original words:

> It is a nation, would I answere *Plato*, that hath no kinde of traffike, no knowledge of Letters, no intelligence of numbers, no name of magistrate, nor of politike superioritie; no use of service, of riches, or of poverty; no contracts, no successions, no dividences, no occupation but idle; no respect of kinred, but common, no apparrell but naturall, no manuring of

---

[5] The following development owes much to a previous article (Lestringant 2004). See, too, Williams 2012, especially 243–8 ('Montaigne and Shakespeare: "Monsieur Monster"'), and Burrow 2016. More generally, see Greenblatt and Platt 2014: 56–71.
[6] On the 'negative formula', see Levin 1970: 113–26.
[7] Seneca 1917–25: XC: 'Men freshly moulded from the hands of the gods.'

lands, no use of wine, corne, or mettle. The very words that import lying, falshood, treason, dissimulation, covetousnes, envie, detraction, and pardon, were never heard-of amongst them. How dissonant would hee finde his imaginary common-wealth from this perfection? (Florio 1999: 1.30)

Reappropriating Florio's translation, Shakespeare provides in turn a clever variant on the same speech:

> Gonzalo: I'th' commonwealth I would by contraries
> Execute all things; for no kind of traffic
> Would I admit, no name of magistrate;
> Letters should not be known; riches, poverty,
> And use of service, none; contract, succession,
> Bourn, bound of land, tilth, vineyard, none;
> No use of metal, corn, or wine, or oil;
> No occupation: all men idle, all,
> And women too, but innocent and pure;
> No sovereignty –
> [. . .]
>
> All things in common nature should produce
> Without sweat or endeavour. Treason, felony,
> Sword, pike, knife, gun, or need of any engine
> Would I not have, but nature should bring forth
> Of it own kind, all foison, all abundance
> To feed my innocent people.
> [. . .]
>
> I would with such perfection govern, sir,
> T'excel the golden age. (2.1.145–66)

The playwright barely transforms Florio's translation but he manages to reinvigorate the original text by dividing it into several speeches. Indeed, Gonzalo's opening lines place the tirade within the Renaissance topos of the world 'upside down', whereas his concluding couplet alludes to the mythological model of the golden age. This double reference – the former being popular while the latter is decidedly erudite – invites a re-evaluation of these borrowed words. For what emerges from this is that Gonzalo's ideal commonwealth oscillates between the humanist dream and the carnivalesque.

Moreover, the dramatic situation characterising the beginning of Act 2, as well as the relationships established between the characters, accentuates the inherent parody of these lines. As he utters them, Gonzalo is lost in a daydream, and he finds himself alone in his reverie. The characters that surround him in the scene either deride his description or become impatient with his ramblings. Sebastian and Antonio at first laugh surreptitiously, then more and more openly, whereas Alonso overtly manifests his anxiety as to the fate of his son Ferdinand, who has not been seen since the storm.

This ironical borrowing of the negative formula, ridiculed by the other characters with the exception of Alonso, who is irritated by his counsellor's naivety, seems to be somewhat of an indirect rebuke to Montaigne, apparently mocked here as a mere idealist. Read in this manner, the ridicule that Gonzalo, the naive and rambling dreamer, receives would function as a posthumous slap in the face to the author of the *Essays*. This analysis therefore implies that Shakespeare was rushed and careless in his own reading, not only cutting and copying from a particular philosophical piece, disregarding context, to develop his character's tirade, but lacerating the text into smaller extracts interspersed with *lazzi*.

For all its coherence, this reading oversimplifies the meaning of Shakespeare's borrowing from Montaigne, and I would like to argue that it is probably not the correct one. Indeed, another interpretation is possible, and it demands a return to Montaigne as a preamble to the reading of Shakespeare. This is the one on which I would now like to focus.

What did Montaigne mean? Or, more so, what is he trying to convey in his use of the 'negative formula'? This structure of negation is in itself highly ambiguous, and found in writers who, like Montaigne apparently, celebrate the survival of the golden age in the South American Indian society, and in those who, to the contrary, appeal to their readers to despise the inability of such men to master the arts and techniques that make modern life bearable. The adepts of 'soft' primitivism rival those supporting 'hard' primitivism; and the unconditional admirers of unclothed, or 'uncivilised', peoples are pitted against the most aggressive supporters of civilisation. Yet all of them share the same rhetorical tool and rely on the 'neither/nor' device. They have the same motto in common, namely this litany of denial, deployed from Gomara[8] to Montaigne

---

[8] See, for instance, Francisco Lopez de Gomara's very successful *Historia General de las Indias* (1552). Gomara's book was translated into English by a Levant merchant named Thomas Nicholas (*The Pleasant History of the Conquest of the West Indias*, London, 1578). A few years later, it was translated into French (*Histoire générale des Indes Occidentales*, II, 96, trans. Martin Fumée, Paris, 1587, ff.188v°–189r°); the chapter entitled 'Des choses nécessaires, desquelles *avaient faute* ces Mexicains' (my emphasis) is particularly worth reading.

and from Cortès to Ronsard. This rhetorical construction entails a great ambivalence, and Shakespeare's manipulation of it is masterful.

Now, does not Montaigne himself play with it in chapter 31 (or chapter 30 in Florio's translation) of the first book of his *Essays*? Only a distracted reader would read into Montaigne a prototype of Gonzalo, at least of the Gonzalo we encounter at the beginning of the play. This unfinished Gonzalo, as it were, is not so different from the comical character of the *pedante* traditionally prattling 'amply and unnecessarily' (2.1.262), exposing his knowledge, or lack thereof, and causing others to laugh at his expense. This can be seen, for instance, in the 'widow Dido' (2.1.75) allusion, echoed and parodied time and again by Antonio and Sebastian, the phrase being at once a comical refrain, a grotesque paronomasia and a sort of silly catchphrase whose sonority intrigues before provoking outbursts of laughter. Nonetheless, Gonzalo will gradually metamorphose into a wise man, a 'Sir Prudence' (2.1.84) that the devilish Antonio and Sebastian will never cease to mock. This 'Sir Prudence' emerges as Prospero's double in some measure, often a ridiculous alter ego of the 'man of science', but definitively more compassionate and humane. Just like Prospero, Gonzalo is indeed capable of insights into life's darker mysteries. Somewhat surprisingly, however, in spite of his extreme lucidity regarding the true essence of human nature, which reveals in him an unexpected pessimism, critics have often perceived him as a naive utopian.

There is nothing to suggest that Montaigne fully adheres to the negative formula. This is rather the contrary, for the text of the *Essays* is more cunning than it first seems, and Shakespeare, as I intend to show, perfectly understands Montaigne's subtleties, which he develops and rewrites in his own way. In the excerpt under study, the negative formula happens to be introduced by a supposition, and this supposition is of prime importance because it puts in perspective the rest of the passage: 'C'est une nation, *dirai-je à Platon*, en laquelle.'[9] In the French text, *dirai-je* (or *diroy je* in the original version) is a future-tense expression with an equally conditional value. Florio makes this clear in his translation and offers his readers an appropriate 'would I answere' (1.30). In a 2001 edition of the *Essays*, the spelling of *diroy je* has been modernised and aptly replaced by

---

[9] Montaigne 1999: 206.
[10] Montaigne 2001: 320.

dirais-je.[10] Montaigne says that he says, or rather that he 'will' or 'would' say, but he does not say anything without precaution. Were that the case, the reader could ascribe directly to him – and without the slightest restriction – the whole series of negations that follow. Critics have already made clear that Montaigne, in his works, constantly and self-consciously plays with the possibilities of language and grammar, and we could see his writings, retrospectively, as the product of someone well versed in 'bathmology', the 'new science' that Roland Barthes neologised for 'degrees of language'.[11] Whether it applies to the *Essays* or to *The Journal of Montaigne's Travels in Italy*, the science of degrees displayed by the French writer allows him to introduce in his text a dialogism between voices within the different layers of the text. In other words, it provides him with a basis for polyphony and encourages infinite reverberations of irony. The discontinuity inherent in theatricality can be read into in the tightly woven textual web of the essays if we use the lens of bathmology.

Shakespeare has clearly perceived the dramatic possibilities contained in Montaigne's *dirai-je* ('would I answere'). Of course, this hypothesis is not in the least plausible, the probabilities of his meeting and quietly exchanging words with Plato but a dream. A humanist, Montaigne revels at the thought of this fictional discussion, which would in essence nullify the twenty intervening centuries of degeneration and forgetting, and, as if by magic, put an end to the gradual loss of intelligence and knowledge that resulted. Ultimately, it is exactly the same nostalgia that grips Gonzalo when he imagines his ideal commonwealth. Consisting in being able to speak on the same level as the Ancients, to debate as peers, or to have a casual conversation as if between equals, friends and contemporaries after such a long absence, this dream was the great longing of the early modern era, and it was definitively lost by Montaigne's and Shakespeare's time.

But since Montaigne was perfectly aware that this dream would never be realised, its mention in the *Essays* is tinged with irony. The hypothesis introduced by the *dirai-je* is irreconcilable with reality and the author knew this well. Assuredly, should Plato be transported to the land of the cannibals, he would be at a total loss. At the very least, he would be puzzled by the fact that Montaigne's commonwealth, contrary to the one he envisaged, could survive with 'so

---

[11] See Barthes 1975, quoted by Laurens 1985. See, too, Force 1997: 187.

little artifice and human patchwork' (1.30). That is indeed a pointed refutation of the *Republic*'s informed and complex architecture.

Gonzalo's Declamation

Allying incongruity to a purely hypothetical issue, this subtle game actually owes much to a well-established rhetorical tradition, that of declamation. More technical than the word 'paradox',[12] that of 'declamation' also has a broader meaning as it designates the practice of oratory development on a given theme as recommended by rhetoricians for the orators' education and training. 'The "real unreal" is indeed the psychological, judicial, and rhetorical object'[13] of the declaimer. In declamation, 'the two essential notions, which are interrelated, are those of exercise and fiction'.[14] Moreover, declamation can be defined by its total freedom, which makes it the privileged instrument of unbiased moral reflexion. Its point of view is fluid; the speaker's identity is continually obfuscated.

Not only because of the instability of the period and of the numerous upheavals that beset it, but also due to its enduring resistance to new ideas, the early modern era turned declamation into one of its favourite modes of expression. Declamation trickles through the whole literature of the sixteenth century, from Erasmus's *Encomium moriae*, first printed in 1511, to Montaigne's *Essays*. In his *Utopia*, Thomas More composed a declamation based on a new geographical model. François Rabelais's work is riddled with declamations – think of Panurge's praise of debts in 'The Third Book of Pantagruel' ('Le tiers-livre de Pantagruel'), chapters 2 to 5, of the hymn on the miraculous herb Pantagruélion, chapters 49 to 52 in the same book, or of the celebration of Gaster, 'prime master of arts in the world', in 'The Fourth Book of Pantagruel' ('Le quart-livre de Pantagruel'), chapters 57 and 61. As to Montaigne's friend, Étienne de La Boétie, he published his *Discours de la servitude volontaire* (*Discourse on Voluntary Servitude*) in 1574, and the whole text can

---

[12] See Dandrey 1997. The perspective I adopt in this chapter roughly corresponds to the one chosen by Dandrey in his study.

[13] 'Le "réel irréel", tel est l'objet psychologique, judiciaire et rhétorique' (Quignard 1990: 15).

[14] '[L]es deux notions essentielles, qui sont liées, sont celles d'exercice et de fiction' (see Chomarat 1981: 935).

be read as a feverish rhetorical declamation marked by an implacable logic.[15]

Not surprisingly, in the *Essays* several chapters, or at least several fragments of chapters, are declamations.[16] One can quote here the praise of kidney stones or *gravelle* in 'Of Experience' (Florio 1999: 3.13), or the juicier plea for the 'indocile libertie' of his virile member in 'On the Force of Imagination' (Florio 1999: 1.20). The best instance of declamation in Montaigne, however, is assuredly the famous essay 'Of Cannibals', a defence of the free anthropophagi of Brazil, the idea of which is revived both in the golden age of the Ancients and the ideal republic envisaged by Plato and Plutarch. Montaigne's characteristic style in his declamation is made of laudatory hyperboles and definitive formulas, like the following one at the very beginning of the essay 'Of Cannibals', which has inspired readings as a relativist slogan: 'men call that barbarisme which is not common to them' (Florio 1999: 1.30). More importantly, it is Montaigne's ironical tone, sometimes playful, sometimes indignant, which generally indicates the use of declamation in his book.

In this regard, the extract from Montaigne's essay on cannibals quoted earlier in this chapter belongs to the trend initiated by Erasmus in his *Encomium moriae*. Indeed, the theme of folly ran throughout early modern literature from Erasmus to Montaigne, and this mild literary folly was neither the harshest nor the most tragic form witnessed by the sixteenth century. Erasmian folly is indeed a gentle, acute madness that implies duplication and theatricality.[17]

---

[15] See Lafond 1984: 736 and Lestringant 1994: 181–3. Montaigne echoes the paradox developed by La Boétie in his *Voluntary Servitude*. Caliban, for instance, becomes the wilful slave of Stephano. In effect, he exchanges his former master for another, even more brutal than the first, replacing the noble Duke of Milan with a drunken butler. From the moment he voluntarily binds himself to this ludicrous and evil character (who, incidentally, also wants to be the king of the island), thereby becoming a slave's slave, Caliban cries out: 'Freedom, high-day! High-day, freedom! Freedom, high-day, freedom!' (2.2.181–2). To which Stephano voices a terse but enlightening comment: 'O brave monster!' (2.2.183). Besides, La Boétie's motif of voluntary servitude is also duplicated in *The Tempest* through an amorous lens. A prisoner of Prospero, the young Ferdinand declares indeed that he wants to be the slave of Miranda, Prospero's daughter, with whom he has fallen deeply in love and whose hand he asks in marriage: 'Ay, with a heart as willing / As bondage o'er freedom. Here's my hand' (3.1.88–9).

[16] See Tournon 1983: 203–28.

[17] See Fumaroli 1972, esp. 95–8.

Incidentally, between the *Encomium moriae* and Montaigne's 'Of Cannibals', one intermediary work can be traced, namely *La Pazzia* ('Folly' in Italian), a work published in Venice in 1540 and translated into French twenty-six years later under the suggestive title of *Les louanges de LA FOLIE, Traicté fort plaisant en forme de Paradoxe* (*The Praise of Folly: A Pleasant Treatise in the Form of a Paradox*).[18] This anonymous pamphlet has been attributed to Ascanio Persio or to Vianezio Albergati.[19] Be that as it may, well before Montaigne, this treatise brought the naked peoples of the New World into the linguistic space of the declamation. In *La Pazzia*, Folly speaks, and she can say everything and allows herself to ruffle the feathers of public opinion. Even more interestingly, she does not hesitate to refute the findings of early modern men of science. Of the people newly discovered in the West Indies, she affirms that they lived happily 'without law, without letters, and without any wise men' ('sans lois, sans lettres, et sans aucuns sages'). Such blissful creatures despised gold and precious jewels ('joyaux précieux'). They knew 'neither avarice nor ambition, nor any other cunning' ('ni l'avarice, ni l'ambition, ni quelque autre art que ce fût'). Taking for food 'the fruits naturally produced by the earth' ('des fruits que la terre sans artifice produisait'), they had 'like in Plato's Republic all things in common, including their wives and infants, for as soon as the latter were born, they fed and brought them up within the community as if they were all theirs'.[20]

His occasional agreement with Plato does not prevent the author from disagreeing with he who would have wanted philosophers to be kings or, at worst, kings to become philosophers. Indeed, *La Pazzia* openly contradicts him: 'On this matter, I will answer no: for the peoples could not suffer a worse distress or calamity than seeing themselves governed by such dabblers in philosophy and excessively

---

[18] Re-edited in 1543: 'Stampata in Venegia per Giovanni Andrea Vavassore detto Guadagnino et Florio fratello.' See its French translation, Thier: 1566. This source has been identified and proposed with a few reservations by Pina Martins 1992; see, too, Pina Martins 1988.

[19] See Dandrey 1997.

[20] See *Les louanges de LA FOLIE*, f.C6v°: 'avaient comme en la République de Platon, toutes choses communes, jusques aux femmes et petits enfants: lesquels dès leur naissance ils nourrissaient et élevaient en communauté comme propres.' Cf. Florio 1999: 1.30: 'Those that are much about one age, doe generally enter-call one another brethren, and such as are younger, they call children, and the aged are esteemed as fathers to all the rest.'

wise men.'[21] It suffices to look at the Spaniards, who 'with their sur-
plus of knowledge, their great contriving, and their harsh and most
unbearable laws and edicts'[22] have filled this land once blessed by
the gods 'with a hundred thousand evils, disputes and ordeals'.[23]
Unsurprisingly, the ease with which the exotic world finds itself
superimposed on the classical is also at work in Montaigne. And it
is Plato, once more, who is the target of this linguistic frivolity: 'It
is a nation, would I answere *Plato*, that hath' (Florio 1999: 1.30).
Such a calculated sense of incongruity carries the mark of *La Pazzia*,
Montaigne simply minimising in his text the somewhat clumsy irony
of his model.

Shakespeare, in turn, interprets and rewrites Montaigne's own
ironical stance, incorporating typically English elements into his
French source. Indeed, his clear borrowing from Montaigne is also a
pretext to return to Erasmus and to More, and to draw much upon
the learned fool or would-be philosopher who wrote *Utopia* (1516),
a text which is nothing other than a long declamation. So, through
Montaigne, Shakespeare revives a tradition initiated by one of his
great predecessors.

Returning to the play, the role of Gonzalo warrants a closer look
in this regard. Alonso, King of Naples, has married his daughter
Claribel in Tunis and he laments his detachment from her. Furthermore, in the storm he lost his only remaining child, his son Ferdinand,
whom he believes to be drowned ('No, no, he's gone'; 2.1.120). It is
to distract him from his mourning and his melancholia that Gonzalo,
without any forethought, launches into his tirade, replete with this
negative formula. As in Montaigne, it is pure supposition with playful intent, whose aim is simply to entertain the king and help him
consign his sad thoughts to oblivion through a series of postulations.
In effect, there is no better way of forgetting about reality and its
harsh cruelty than to forge a fiction as remote as possible from any
semblance of it. Logically, then, this is precisely what Gonzalo does.
Of course, his ideal State does not have the slightest chance of ever
being realised:

---

[21] Ibid., f.C7r°: 'Là-dessus je répondrai que non: mais que les peuples ne sauraient
être plus malheureux, ni en plus grande calamité, que de se voir tomber ès mains
de tels philosophâtres et trop sages hommes.'

[22] Ibid.: 'avec leur trop de savoir, leurs grandes finesses, leurs très dures et insup-
portables lois et edits'.

[23] Ibid.: 'de cent mille maux, fâcheries et travaux'.

Gonzalo (to Alonso): Had I plantation of this isle, my lord –
Antonio: He'd sow't with nettle-seed.
Sebastian: Or docks, or mallows.
Gonzalo: – And were the king on't, what would I do? (2.1.141–3)

This image is a far cry from reality. Gonzalo is not, and will never be, a king; he is fully aware of this and he does not aspire to become one. Contrary to Antonio and Sebastian, his contradictors, he does not aim at usurping Alonso's throne. He actually saves his prince's life by waking him up just in time when he realises that the two traitors have drawn their swords to kill him in his sleep (2.1.305–6). So, if Gonzalo is king, for the brief moment of his reverie, he is but a mock or carnivalesque king, harmless and benevolent, who abdicates all sovereignty for the benefit of his subjects.

Moreover, Gonzalo's unbelievable hypotheses are never really coherent. The old man mixes up several models which hardly fit together: that of the world turned upside down mentioned at the beginning of his tirade ('I would by contraries / Execute all things'; 2.1.45–6), that of the golden age mentioned in closing (2.1.166), and, last but not least, that of the ideal republic informing his entire speech. It seems that, in the sophisticated mode of erudite declamation, Gonzalo plays the king's fool. He is a noble and gentle jester, so to speak, who stands in sharp contrast to Trinculo, the common clown.

On top of this, like Folly in Erasmus's declamation and Montaigne in his *Essays*,[24] Gonzalo contradicts himself and does not seem to be aware of it. Antonio and Sebastian have good reason to notice the discrepancy between the first and the last lines of his speech. Indeed, Gonzalo would like to be king ('And were the king on't'; 2.1.143) of a people who have no notion of sovereignty ('No sovereignty'; 2.1.154), so much so that, as Antonio puts it, 'The latter end of [Gonzalo's] commonwealth forgets the beginning' (2.1.155).

The contradictions were no less blatant in 'Of Cannibals', where, in a way, we witness a similar configuration. Montaigne begins by telling us that cannibals ignore all the inventions, and therefore all the evils, that characterise our society. In his essay, the cannibals – in reference to Rio de Janeiro's Tupinikin Indians, a tribe whose ferocious feats had

---

[24] See Florio 1999: 3.2 ('Of Repenting'): 'Howsoever, I may perhaps gaine-say my selfe, but truth (as *Demades* said) I never gaine-say.'

been divulged by André Thevet and Jean de Léry a few years before –
are seen as an ideal type complementing all the poetic clichés generally
used to depict the golden age ('all the pictures wherewith licentious
Poesie hath proudly imbellished the golden age'; Florio 1999: 1.30).
There is nothing new, in fact, in the litany of negation resorted to by
Montaigne and later adopted by Gonzalo.

Yet Montaigne is well aware of the fact that, in reality, these Indians practise agriculture, spin and weave cotton, have a bartering
economy, an extremely complex system of parentage, and continually wage war against one another. As a matter of fact, in the rest of
the chapter, he seems to forget the initial negation. At the risk of dis-
appointing the primitivist's imagination, Montaigne then proceeds
to construct a positive figure of the Brazilian, giving details on all
the material circumstances of their existence: their hammocks and
weapons, their food and drink made 'of a certaine root, and of the
colour of our Claret wines' (ibid.), the staff they use to drum ('great
Canes open at one end, by the sound of which they keepe time and
cadence in their dancing'; ibid.), but also their gestures and manners,
their music, and their love poetry worthy of the Anacreontic metre
('this invention hath no barbarisme at all in it, but is altogether Ana-
creontike'; ibid.).

To put it differently, the primitivist topos is merely a phase in
Montaigne's reasoning, just like the allusion to Lycurgus's and Plato's
ideal republic. The negative formula represents the *tabula rasa*, a
blank slate on which an anthropological reconstruction becomes possible. Montaigne only employs this frame of reference to subvert and
work around it. It could seem as if he sets up ideas just to shoot
them down one by one in his disposal of established thought patterns, topoi, myths, sentences and ancient fables such as those from
the golden age, the time of Atlantis, the ideal republic or the naked
philosophers. But he does not do so gratuitously. That Montaigne
scatters topoi in his text to better shape his thoughts should not lead
us to conclude that he is simply taking pleasure in exposing false-
hoods – in order to *deconstruct* them, as we would put it today. We
can distinguish a real search for a solid foundation of meaning in
these shifts and the continuous journey that thought undergoes. In the
meantime, the intellect is on unstable ground, loose soil or a storm-
tossed sea, on a fickle island peopled with voices and evil spirits.

In *The Tempest*, the ingenious fiction which Gonzalo recounts
only elicits Alonso's annoyance. The king's weariness finds no comfort in his counsellor's declamation; indeed, to the contrary, he

orders: 'Prithee no more. Thou dost talk *nothing* to me' (2.1.170, my emphasis). This 'nothing' could actually be the foundational definition of declamation – declamation being a 'nothing' that opens up and deploys all the possibilities of meaning.

Knowledge and Amazement

Gonzalo is in a constant state of amazement throughout the play. The crew has barely landed on the island when he starts rejoicing about their miraculous 'preservation' (2.1.7), manifesting his happiness to the point of upsetting Alonso, who is in mourning over his son Ferdinand, lost and presumably drowned. Gonzalo then marvels at the island's fertility: 'Here is everything advantageous to life' (2.1.50). This observation is in all likelihood an echo of Montaigne's 'Furthermore, they live in a country of so exceeding pleasant and temperate situation, that as my testimonies have told me, it is verie rare to see a sicke body amongst them' (Florio 1999: 1.30). Not seeing the 'lush' (2.1.53) paradise mentioned by the counsellor, Antonio and Sebastian's reaction to Gonzalo's exclamation is to sneer, as they can only make out swamps and heath.

Their clothing, still intact and unblemished, is the cause of Gonzalo's third astonishment. Indeed, the clothes worn by the crew have been 'drenched in the sea' and yet they surprisingly retain 'their freshness and gloss' (2.1.61–3), just as they were at 'the marriage of the King's fair daughter Claribel to the King of Tunis' (2.1.69–70). Which leads to the fourth: he recollects that Tunis is built on the site of the ancient city of Carthage, and Claribel's wedding reminds him of the mythical love affair between Aeneas and Dido, the 'widow Dido' celebrated by Virgil and Ovid. But the parallel is somewhat awkward, since it is a story of a glorious love affair but one that ended in tragedy, thus highly unlikely to bring any solace to Alonso in his distress. This outburst of astonishment shows the incongruity of Gonzalo's character at its peak, and yet it is perhaps the key to understanding his character. He believes in the power of words, which is clearly not the case for his two contradictors, two dark souls who are sarcastic about miracles and legends: 'His word is more than the miraculous harp' (2.1.85). Hence their outpouring of mockery, comparing the island's metamorphosis into an apple (2.1.88–9) and Gonzalo's fantastical monologues to 'kernels' thrown in the sea and generating a fabulous archipelago ('bring forth more islands'; 2.1.90–1).

Yet the old man's series of astonishments marks a turning point in the plot. Gonzalo suddenly comes back to reality and protests against Sebastian as the latter blames Alonso for having married his daughter to 'an African' (2.1.123) in Tunis rather than giving her to a European prince. Having purportedly lost his daughter before losing his son, Alonso would be the cause of all the crew's misfortunes, if one were to believe his brother. Gonzalo is outraged by an accusation that, if not totally ungrounded, is lacking in courtesy: 'The truth you speak doth lack some gentleness' (2.1.135). The counsellor, in himself, is deprived neither of compassion nor of gentleness (or *gentillesse*, as Montaigne would have put it), i.e. of nobility.

Thereafter, Gonzalo's true nature is gradually revealed and we come to realise that his fondness for marvels and ancient fables is not the product of an erudite struck dumb with admiration. Inspiring sarcastic remarks from the villains, his fantastical evocations actually help him to put up with the cruelty of existence: even if they are sometimes off-target, they have their compensatory value. It is precisely when Gonzalo resorts to them that, apparently out of the blue, he offers a pastiche of Montaigne in borrowing the litany of negation from 'Of Cannibals'. Ultimately, Gonzalo is no more deceived than Montaigne by the prestige of fables, but he similarly believes that fables are necessary for our infirmities, and, so to speak, consubstantial with our humanity. He is a humanist at heart but has no illusions on man's real nature; his position could be exemplified by Prospero's famous meditation according to which 'We are such stuff / As dreams are made on' (4.1.157–8).

Gonzalo has a certain lucidity: as mentioned above, it is he, inspired by Ariel's singing voice, who wakes up the sleeping king ('Now, good angels / Preserve the King!'; 2.1.304–5). At the end of the scene, the double entendre inherent in his pronouncements suggests that he has perceived wherein lies the danger – not from a lion nor even from 'a whole herd of lions' (2.1.314), but from men much worse than savage beasts. In an exchange with Alonso, intent on resuming the search '[f]or [his] poor son', he exclaims: 'Heavens keep him from these beasts!' (2.1.322). Gonzalo's character is no longer ridiculous from this point. He even becomes the object of Alonso's sympathy, when in his exhausted and aching state he requests some rest: the king, also afflicted with tiredness and pain, responds, 'Old lord, I cannot blame thee' (3.3.4).

Then, when the celestial music orchestrated by Prospero resounds, once more prompting Gonzalo's amazement ('Marvellous sweet music!'; 3.3.19), the counsellor proffers another double-edged observation that Prospero, present but invisible, aptly decodes. Indeed, he observes that the monsters inhabiting the island have in fact 'manners [. . .] more gentle-kind' (3.3.32) than the majority of men. One can clearly see here how far removed Gonzalo is from the naive optimism we could have hastily ascribed to him at first sight. At the same time that he is apparently the most idealistic of the ship-wrecked lords, he is also the most clear-sighted. And while, just like Montaigne's essay on the cannibals, his political fable is seemingly innocent, it turns out to be much more relevant than it seems, as the unfolding of the plot will show. For Gonzalo's capacity for amaze-ment does not preclude an acute critical insight.

No wonder, then, if his amazement quickly ceases to be an object of ridicule in the play. When confronted with the strange shapes con-jured up by Prospero's magic, he imagines himself to be one of those '[t]ravellers' (3.3.26) reporting extravagant wonders from their voy-ages at sea, and we feel that, more than the other characters, he deserves to be listened to and taken at his word. In this regard, it is worth noting that, in England, belief in the geographical won-ders inherited from antiquity and the Middle Ages had been hold-ing firm ground for a longer period than anywhere else in Europe. For instance, Mandeville's travels remained extremely popular until Shakespeare's time, with the apparition of Richard Hakluyt's 1589 *Principall Navigations*,[25] edited in English, and Sir Walter Raleigh's 1599 *Brevis et admiranda descriptio regni Guianae*, which still, at the very end of the sixteenth century, depicted Amazons with bows and arrows and offered side and back views of Blemmyes who had their faces on their chests.[26] Engravings such as those illustrating Raleigh's description of Guyana are probably the direct sources of Gonzalo's allusion to men '[w]hose heads stood in their breasts' (3.3.47). But the credulity of the 'old lord' should not be assessed from our current perspective. Indeed, his capacity for amazement, untouched by his extensive reading, matched that of many of his early modern con-temporaries. Incidentally, in Act 3, scene 3 he is preceded by Alonso, Sebastian and Antonio, who are similarly surprised and delighted by

---

[25] Hakluyt 1589: 25–77: 'The voyage of John Mandevill knight in latin, begun in the raigne of Edward the 2. Anno 1322, continued for the space of 33. yeeres, and ended in the raigne of Edward the 3. from England to Iudea and from thence to India, China, Tartaria, and as farre as 33 degrees to the south of the Aequinoctiall.'

[26] See Raleigh 1599: pl. 15. The original narrative was published just a few years earlier under the following title: *The discoverie of the large, rich, and bewtiful empyre of Guiana* (1596).

the extraordinary and incomprehensible scene taking place before their eyes. And they, too, momentarily abdicate all critical insight as they discover these 'several strange shapes' (3.3.19, s.d.).

Gonzalo's Books

The Tempest therefore presents us with a world of books posed between science and superstition, fact and fiction, and questioning the nature of early modern knowledge. From his own books, Prospero draws his supernatural powers and, as such, a sort of super-humanity. It is through his books that he can harness nature's forces and even seize Jupiter's thunder. Significantly, as he prepares himself to leave the island of all enchantments in order to return to Europe and Milan, he yields up his powerful magic and throws his book into the sea: 'I'll drown my book' (5.1.57).

Books actually draw Gonzalo and Prospero closer. We learn how the 'noble Neapolitan' (1.2.161) helped Prospero bring his books with him when the latter was removed from his dukedom and how, to a certain extent, he saved Prospero's books ('he furnished me / From mine own library with volumes that / I prize above my dukedom'; 1.2.166–8), those from which Prospero draws his power over nature and men. It is not by chance that Gonzalo shares adversaries with Prospero, nor that the malevolent characters of Antonio and Sebastian – and, to a lesser extent, Alonso – mock Gonzalo and his bookish references. They are the enemies of books, hence of intelligence and humanity in its most dignified aspects.

Like Prospero and Montaigne, Gonzalo is an assiduous reader. But in *The Tempest* we need to consider genre: there are of course *grimoires*, or spell books, that enable their readers to command natural elements and human beings, but also story books, be they fictional or historical, whose over-consultation can engender a kind of maladjustment to the real and the present. If the all-powerful Prospero draws his strength from his books of magic, it seems that Gonzalo, to the contrary, is left with a form of weakness or even impotency inherent in contemplative life – or in scholarship, as we would put it today.

In fact, it is all of Gonzalo's *humanitas* that can be said to originate from his books. His humanity is anchored in the history and in the natural philosophy of classical writers such as Virgil and Pliny, whom he does not name but whom he has read and seems to recite from time to time. Such readings account for his admiration of 'widow Dido' and his belief in the existence of the monstrous races described by Pliny the Elder and early modern cosmographers like Mandeville, Hakluyt and Raleigh.

While Prospero attains the inaccessible ideal of any intellectual, which consists in combining action with knowledge, Gonzalo can only bear witness to their painful and risible divide. This is precisely what likens him to the reader and to Montaigne, and what also tends to make him a ridiculous figure at one moment, pathetic the next. It is the situation of the humanist confronted with a barbaric world. For what can we expect of a poor humanity faced with 'beasts' such as Antonio and Sebastian that have nothing human about them but their faces? The cultural memory of antiquity, in all its status and prestige, seems to lack here in their being able to provide lessons or remedies for the present.

If one is to believe Claude Lévi-Strauss in *The Story of Lynx*, such is the question running through the *Essays* and, in particular, through the 'Apologie of Raymond Sebond' (Florio 1999: 2.12):

> Montaigne's philosophy states that any certainty has the a priori form of a contradiction and that there is nothing to seek underneath it. Knowledge and action are forever placed in an awkward position: they are stuck between two mutually exclusive systems of reference that are imposed on them though even temporary faith in one destroys the validity of the other. And yet we need to tame them so that they can live together in each of us without too much drama. Life is short; it is just a matter of a little patience. The wise person finds intellectual and moral hygiene in the lucid management of this schizophrenia. (Levi-Strauss 1996: 217)[27]

So, how to bridge the gap between knowledge and action? While Montaigne, philosopher and moralist, makes an assessment, Shakespeare, man of the theatre, suggests two solutions. These are magic, i.e. drama or theatre, and irony, which presupposes distance, performance and duality, i.e. drama, once again.

---

[27] See Lévi-Strauss 1991: 288: 'La philosophie de Montaigne pose que toute certitude a la forme *a priori* d'une contradiction, et qu'il n'y a rien à chercher par-dessous. La connaissance, l'action sont à jamais placées dans une situation fausse: prises entre deux systèmes de référence mutuellement exclusifs et qui s'imposent à elles, bien que la confiance même temporaire faite à l'un détruise la validité de l'autre. Il nous faut pourtant les apprivoiser pour qu'ils cohabitent en chacun de nous sans trop de drames. La vie est courte: c'est l'affaire d'un peu de patience. Le sage trouve son hygiène intellectuelle et morale dans la gestion lucide de cette schizophrénie.'

The magic of Prospero, an exceptional character who resorts to exceptional means, depends on the power of books, or rather of one particular book, and on what they tell us about the world. Yet, even in this specific case, one should reassess the value of such a book, for the world in which Prospero lives and which he dominates, thanks to his *grimoire*, is circumscribed within the space of an island: a closed and well-delimited space like the defined stage of a theatre. 'The island is the metaphor of the stage and the stage is the projection of the universe,' Jean-Michel Racault writes, and, as he puts it, '[t]he architectural layout of Elizabethan theatres – circular theatres surrounded by galleries – contributes to reinforce this analogy' (2010: 91).[28]

Prospero's island is thus both the small-scale model and the simulacrum of the world, a world in miniature on which the mind, through the medium of books, can anchor itself, contrary to the real world, whose immensity and depth are beyond its grasp. That is the reason why Shakespeare has replaced the 'infinit and vast' country (Florio 1999: 1.30), or *'païs infini'*, of Montaigne's cannibals with Prospero's island, a topographical object more easily controlled by a superior and benevolent intelligence.

Far from being a *tabula rasa*, *The Tempest* has its grey areas, its forests and its swamps. It is where the witch Sycorax gave birth to the black Caliban before being buried there, with all her charms. Two mischievous rascals, namely Antonio and Sebastian, then land there, together with a couple of drunkards, Stephano and Trinculo. Undoubtedly, evil is rampant on the island. It inhabits it. But it does not yet possess it entirely, and forgiveness will triumph in the end.

Magic and drama will take their leave once the characters depart from this island of dreams and reach the solid ground of reality. Then, they will have to cope with it: through mercy, through forgiveness, through irony. They will have to get by as if dreams, this stuff we are made of, were not incompatible with the real and could mould themselves into its shape. As the denouement draws near, Prospero's words of forgiveness to his vanquished adversaries, whose villainy has amply been demonstrated and confirmed throughout the play, are a clear-headed means to reconcile an ideal conception of humanity with the deeply inhuman realities that lie dormant within us. Gonzalo will continue to find amazement in the world, even though he knows all too well that his awe can be refuted and contradicted at every moment, engulfed in the bottomless abysses of a harsh reality.

---

[28] More generally, see Racault 2010: 63–92. See also Gardette 1994.

The same can be said for Montaigne, for whom 'the novelty of things doth more incite us to search out the causes, than their greatness: we must judge of this infinit powere of nature, with more reverence, and with more acknowledgement of our owne ignorance and weaknesse' (Florio 1999: 1.26). In the vein of St Augustine, Montaigne gives back their real, primary meaning to miracles, as they are linked to man's amazement faced with the disparity of the world and the self:

> I have seene no such monster or more expresse wonder in this world than my selfe. With time and custome a man doth acquaint and enure himselfe to all strangenesse: But the more I frequent and know my selfe the more my deformitie astonieth me, and the lesse I understand my selfe. (Florio 1999: 3.11)

The concrete man emerges from the ruins of dogmatic pride in all his complexity, as a source of wonder.

Conclusion: The 'capacity for wonderment'

According to Jean Céard in *La Nature et les prodiges*, Montaigne thus invites us to 'disaccustom ourselves of who we are in order to awaken in us the capacity for wonderment'.[29] Is not Iris, the messenger of the gods, 'the daughter of Thaumantis', i.e. of the centaur Thaumas whose very name is derived from the Greek verb 'to admire'? Montaigne adds: 'Admiration is the ground of all Philosophy; Inquisition the progresse; Ignorance the end' (Florio 1999: 3.11).

This may allow us to understand Antonio and Sebastian's ultimate conversion to miracles and wonders. Indeed, after having so long derided Gonzalo, they eventually come to share his capacity for awe. When, in Act 3, Prospero conjures up before them, with Ariel's help, a performance of spirits in the guise of a puppet show, they all become enthralled by what they see. Sebastian exclaims:

> Now I will believe
> That there are unicorns; that in Arabia
> There is one tree, the phoenix' throne, one phoenix
> At this hour reigning there. (3.3.21–4)

---

[29] Céard 1977: 426: 'nous désaccoutumer de nous-même et à réveiller en nous l'aptitude à l'étonnement'.

In this, he is echoed by Antonio in the following lines:

> I'll believe both;
> And what does else want credit, come to me,
> And I'll be sworn 'tis true. Travellers ne'er did lie,
> Though fools at home condemn 'em. (3.3.24–7)

From this point on, they are saved – and thereby forgiven. To some extent, Prospero won them over to Gonzalo's cause, which was first and foremost his own. This cause is ultimately that of books, which, thanks to the fables they contain, open up the inexhaustible world of imagination and can make life both understandable and bearable, even if it is neither the most direct nor the easiest way of accomplishing such a feat.

The cause for books, in *The Tempest*, is the cause for one book in particular. Gonzalo's book is Montaigne: *Montaigne's Essays*, or *Gonzalo's Books*.

Chapter 9

Unlimited Science: The Endless Transformation of Nature in Bacon and Shakespeare's *The Tempest*

Mickaël Popelard

Shortly before joining her husband in 'doing the deed' (*Macbeth*, 2.1.14), Lady Macbeth rejects his orthodox conception of human nature. Whereas Macbeth argues that '[he] dare[s] do all that may become a man; / Who dares do more is none' (1.7.46–7), Lady Macbeth holds the contrary view that the essence of man lies in infinity, which she conceives as the quality or condition of lacking bounds or limits: 'When you durst do it, then you were a man; / And to be more than what you were, you would / Be so much more the man' (1.7.49–51). According to Lady Macbeth, the more you do, the more human you are: in other words, pushing back the boundaries of human nature is not akin to transgressing the limits which make you what you are (i.e. a human being). On the contrary, only by extending those limits can one fully realise one's human essence. A craving for the unlimited is what defines humankind, or in the words of Terry Eagleton, in his follow-up to *Literary Theory*,

[i]t is a quarrel between those like Macbeth who see the constraints of human nature as creative ones, and those like Lady Macbeth for whom being human is a matter of perpetually going beyond them. For Macbeth himself, to overreach those creative constraints is to undo yourself, becoming nothing in the act of seeking to be all. It is what the ancient Greeks knew as *hubris*. For Lady Macbeth, there is no such constraining nature: humanity is free to invent and reinvent itself at will, in a potentially endless process. The more you do, the more you are. For his part, Aristotle would have sided with Macbeth. He thought that the idea of economic production for profit was unnatural, since it involved a boundlessness which is alien to us. (Eagleton 2004: 118–19)

The idea that the Renaissance, as a historical period, developed an interest in the twin notions of 'infinity' and 'boundlessness' is certainly not new. Historians of science have shown that the shift from a closed world to an infinite universe was one of the major intellectual events defining the 'scientific revolution' (Johnson and Larkey 1934; Koyré 1957; Johnson 1968). Together with the mathematisation of nature, the substitution of an indefinite or even infinite universe for the old hierarchical and well-ordered cosmos of the Greek and medieval thinkers represented one of the two major philosophical changes brought about by the intellectual and scientific 'revolution' of the seventeenth century.[1]

In this chapter, however, I am not primarily concerned with the cosmological hypothesis of the infinity of the universe. Instead, I would like to suggest that Lady Macbeth's iconoclastic idea of human nature as a plastic and malleable reality which can be kneaded into different shapes is in line with a new conception of the *physical* world which permeates both Bacon's philosophy and Shakespeare's *The Tempest*, albeit to different degrees. Much as Lady Macbeth argues for the unfixedness of human nature, Bacon also insists that the physical world is by definition changeable and liable to improvement. If Macbeth is encouraged by his wife to become a stronger, bolder and more ambitious man, the natural philosopher is entrusted by Bacon with the task of transforming nature into a richer, more varied and ultimately better place. As the Father of the House of Salomon explains in *New Atlantis*, the aim of his foundation is '*the enlarging of the bounds of human empire*, to the effecting of all things possible' (Bacon 1999: 177, my emphasis). The boundlessness of nature, therefore, is not so much discovered or encountered in a haphazard way as it is methodically 'invented'[2] or

---

[1] See Koyré 1957: viii: 'the changes brought forth by the revolution of the seventeenth century [. . .] seemed to me to be reducible to two fundamental and closely connected actions that I characterised as the destruction of the cosmos and the geometrization of space, that is the substitution for the conception of the world as a finite and well-ordered whole [. . .] of that of an indefinite or even infinite universe'.

[2] 'Invention' has become a very popular notion with early modern scholars, as Fleming notes in *The Invention of Discovery, 1500–1700*: '*Invention*, as opposed to discovery, is now the academic leitmotif of the early-modern [. . .] Far from merely finding new aspects of the world as it was – it is now argued – the expansionist Europe of the sixteenth and seventeenth centuries *made* its new worlds, in many aspects, as these have come to be' (Fleming 2011: 1). See, for some other illustrations of this 'academic leitmotif', Judith Veronica Field's *The Invention of Infinity: Mathematics and Art in the Renaissance* (Field 1997) and David Wootton's more recent *The Invention of Science: A New History of the Scientific Revolution* (Wootton 2015).

constructed. Unlike the astronomer who cannot be said to be instrumental in actually *creating* the infinity of the universe,[3] the natural philosopher plays an active role in endlessly transforming his physical environment. Far from being a mere 'fact', the boundlessness of nature thus becomes the theoretical construct and practical consequence of the new science which Bacon theorises and which Shakespeare stages in *The Tempest*. What I hope to show is that the scientific and philosophical shift from closure to openness – or from a fixed nature to a boundless and plastic one – which characterises Bacon's new scientific method may also be said to critically resonate with Shakespeare's *The Tempest*, albeit in a more circumspect way.

Francis Bacon's Open Reform of Science and the Endless Transformation of Nature

According to Fleming, the shift from pre-scientific to scientific natural philosophy was accompanied by a new hermeneutics of discovery whereby scientists and philosophers alike tended to substitute allegory for literalism in the interpretation of nature. Yet it was important for allegory not to turn into an infinite quest for meaning and truth, for 'the way things truly are, insofar as it can be grasped or known, must then be protected from further interpretative penetration, lest scientific factuality produce infinite regress' (Fleming 2011: 6). Fleming goes on to illustrate his point by citing Galileo and Newton who, he writes, were anxious to set limits on the interpretative process and adopted a 'this-far-but-no-further' (ibid.) principle as the rule of their hermeneutics. The notion that the philosopher should only 'go this far but no further' is one that appears profoundly at odds with Baconian philosophy, especially if one chooses to take Bacon's voyaging analogy seriously. In Bacon's philosophy, as I will endeavour to show, infinite theoretical and practical progress takes precedence over Fleming's pessimistic vision of infinite (interpretative) regress. The engraved title of *Instauratio magna* (1620) represents the progress of knowledge in the form of a ship sailing past the pillars of Hercules, while the motto ('multi pertransibunt et augebitur scientia') is an invitation to the philosophically inclined reader to 'go further' so that *scientia* may be increased. By substituting Daniel's prophecy about

---

[3] Except, perhaps, in the sense that infinity must first be 'theorised' before it can be 'seen' (in a manner of speaking), since one only sees what one is prepared to see.

the progress of knowledge (Daniel 12:4) for the 'non plus ultra' monitory inscription which the pillars of Hercules were supposed to bear,[4] Bacon is, in effect, calling for a removal of interpretative boundaries. The closed and lesser seas of the old logic are supplanted by the unbounded waters of Bacon's *Novum organum* (1620). His new method is like the mariner's compass since it will help men to 'cross the oceans and discover the regions of the New World' instead of sailing by the stars and '[coasting] along the shores of the Old World' (Bacon 2004: 18) or 'crossing the mediterraean sea' only (Bacon 2004: 19). As he himself explains in the preface to the *Novum organum*, a 'this-far-but-no-further' principle may therefore play a detrimental role by stifling investigation. It may even lead to a form of arrogant dogmatism (Bacon 2004: 53).

By contrast, Bacon advocates an 'open-ended philosophy'. He has, he says, 'no universal or systematic theory to put forward' (Bacon 2004: 175). What he does put forward, however, is a new method for acquiring knowledge of, and power over, the natural world. In Bacon's mind, the true approach to the understanding of nature, or 'Interpretation of Nature', ascends gradually and by degrees from the 'sense and particulars' to true 'forms'. By discovering the 'form' of a given nature, one acquires both true knowledge (what Bacon calls *contemplatio vera* in *NO*, II, 3) and 'unrestricted operation' (or *operatio libera*). But nowhere does Bacon claim that his method is complete.

On the contrary, he makes it quite clear that the task he has set himself is too colossal for a single mind to undertake, however bright and penetrating such a mind might be. As Rees wisely remarks, 'Bacon by contrast offered an open philosophy, one which was as incomplete as it was new, and one in which incompleteness was part of its novelty' (Rees in Bacon 2004: xlv). Most of Bacon's major works – including the *Novum organum* and the *New Atlantis* (1627) – are unfinished and they are addressed to posterity. Bacon's philosophy, therefore, may be seen as a work in progress, a collective undertaking every reader is invited to participate in. It is therefore necessarily 'indefinitely new and indefinitely incomplete' (Bacon 2004: xlv), for completeness – which translates into the *magnalia naturae* or wonders of nature, the list of which is appended to *New Atlantis* and includes,

---

[4] As Earl Rosenthal remarked, it wasn't until the sixteenth and seventeenth centuries that the motto acquired a monitory significance. In Charles V's time, however, the columns had already started to become Promethean symbols. See Rees's *Commentary on IM Preliminaries* (Bacon 2004: 490) and Earl Rosenthal's 'Plus ultra, non plus ultra, and the Columnar Device of Emperor Charles V' (Rosenthal 1971).

among many other similar goals, 'the curing of diseases accounted incurable' and the 'restoration of youth' (Bacon 1999: 185–6) – can only be the unreachable horizon towards which science is moving, slowly but surely. In the words of Graham Rees again, 'this is a future of steady, cooperative progress towards completeness, to a promised land of material plenty, and freedom from physical affliction' (Bacon 2004: xlv). That Bacon saw himself as laying the groundwork for a collective and intergenerational philosophical enterprise is made manifest at the end of the first book of the *Novum organum*: 'I do not affirm that nothing can be added to what I prescribe; on the contrary, as one who observes the mind not only in its innate capacity but also insofar as it gets to grips with things, it is my conviction that the art of discovering will grow as the number of things discovered grows' (*NO*, I, 130, in Bacon 2004: 197).

As can be inferred from these lines, Bacon implies that the true interpretation of nature will continue to develop like some monstrously gigantic tree – an apt simile, perhaps, if one considers, as Brian Vickers pointed out in his study of Bacon's prose, that 'the greatest of the image-groups in Bacon's writing [is] that of natural growth' and that 'a tree and its cultivation is also an image for intellectual growth in general terms' (Vickers 1968: 193–4). Bacon draws a clear distinction between his own plan for the reform of science and what the existing schools of philosophy have promised so far, which, he says, is unattainable and unrealistic, because their theories are based on sand – or, to be more precise, on 'impetuous and premature' reasoning (*NO*, I, 26, in Bacon 2004: 75) – thus rejecting what he sees as the three main kinds of false philosophy, namely the sophistical, the empirical and the superstitious, each of them encapsulating a different yet equally flawed way of reasoning about nature.

In contradistinction to the 'false' philosophers, the 'true' philosophers must steer a middle course between dry rationalism and blind empiricism. Like the bee, they must collect their material from the natural world before digesting it: true philosophy therefore consists in using the powers of the mind to collect, analyse and process the data gathered from natural history. Drawing from such an appropriate empirical basis and following a new kind of induction based on the principles of exclusion and rejection, the mind then gradually ascends to true axioms before descending into 'works'. A better – i.e. truer – method will yield unprecedented results, which is why Bacon believes that there is reason for hope: 'therefore from a closer and purer alliance (not so far achieved) of these two faculties (the experimental and the rational) we should have good hopes' (*NO*, I, 95, in Bacon 2004: 153). Such a root-and-branch reform of philosophy may take time but Bacon is confident that it is not doomed to failure. In fact, endless though it may seem, there is every reason to believe that it can, and possibly will, be successful:

> But while this may seem to be an endless task from the outset, and beyond the powers of mere mortals, yet when taken in hand, it may be found to be more reasonable and moderate than those proposals acted on hitherto. For this matter can come to a conclusion whereas the proposals already implemented in the sciences leave men forever spinning in dizzying circles. (*Great Instauration Preliminaries*, in Bacon 2004: 5)

Bacon does not say that his reform *will* come to a conclusion, only that it *can* come to a conclusion. Progress towards completeness will be steady but necessarily asymptotic. Elsewhere, he also explains that he has played the role of a 'trailblazer following in no man's footsteps', yet 'setting out steadfastly on the right road' in his quest for truth (*NO*, I, 113, in Bacon 2004: 171). His path, he continues, 'is, unlike that of the rationalists, open not only to single travellers but one in which men's work and labour (especially for the gathering in of experience) can best be shared out and then brought together' (Bacon 2004: 171). Collaboration and the passage of time are thus two essential aspects of Bacon's reform of philosophy, and they both imply open-endedness, for one more fellow worker and some more time are only likely to make it more perfect than it already is. Thus, by pitting the image of 'the right road' against that of the 'dizzying circles' of received philosophies, Bacon unwittingly reinforces the notion that his reform of science is necessarily incomplete and limitless. To take a mathematical simile, Bacon's open-ended reform of philosophy is like a straight line stretching off into infinity so that the interpretation of nature becomes a distant goal towards which one can only strive asymptotically, whereas the existing schools of philosophy are like closed circles endlessly repeating themselves but leading nowhere near the true understanding of the natural world.

Bacon's Idea of Science and the Endless 'Perfecting' of Nature

Bacon's reform of science is therefore 'open-ended' by virtue of its collaborative and intergenerational nature. More importantly, perhaps, it is also conducive to boundlessness because it entails the never-ending transformation of nature. Although Bacon's world is finite,[5] it is also marked by continuous change as a result of man constantly and relentlessly transforming it. As will be remembered, one of Bacon's key concepts is that of 'form'. But despite the distinctively Aristotelian or scholastic ring of the concept, Bacon's forms are very different from their Greek or medieval predecessors. In fact, as Perez-Ramos has shown, 'Bacon does not conceive his Forms as ultimate units of contemplative grasp' (Perez-Ramos 1988: 86).[6] On the contrary, the forms may be said to bridge the gap between 'speculation' and 'operation', for, as Bacon explains in the *Novum organum*, 'from the discovery of forms flows true speculation and unrestricted operation' (*NO*, II, 3, in Bacon 2004: 203). In Baconian thought, therefore, boundlessness takes the form of the 'unrestricted' or 'unlimited' transformation of nature. By placing such a premium on operation and making it inseparable from speculation, Bacon presents knowledge and power as the twin aims of natural philosophy, thus stating the convertibility of the speculative form into an operative rule. But, in so doing, Bacon markedly departs from the Aristotelian tradition of *scientia* as 'right talking'.[7] As Perez-Ramos explains, Bacon's celebrated equation between *scientia* and *potentia* is completely novel in the philosophical literature of his age:

> Nowhere in the Aristotelian tradition is the term *scientia* so expressly linked with a purposive attempt to alter the course of Nature's processes so as to denote the field of operative knowledge. Man is or can become sapiens or 'knower' as a self-perfecting beholder, not as a self-debased maker [. . .] Indeed, it was generally taken for granted that the chief function of scientifically informed language was to articulate the results of perception in accordance with the accepted principles of explanation. (Perez-Ramos 1988: 85)

Therefore, in the Aristotelian and scholastic philosophical traditions, for a given thing to become an object of science, it must be embedded in a proper linguistic mould: scientific knowledge is defined in

---

[5] Bacon was not a supporter of heliocentricism and he believed that the universe was finite, with the earth sitting at its centre. See Rees, 'Introduction', in Bacon 2004: lxxxiv.

[6] On the importance of Perez-Ramos's book for our understanding of Bacon's philosophy, see Vickers 1992.

[7] As noted by Perez-Ramos, this definition of Aristotelian science is borrowed from J. H. Randall (Randall 1962: 32–58).

linguistic, not practical, terms. Conversely, for Bacon, science acquires a distinctively practical quality. No longer exclusively verbally oriented, Bacon's science may be said to substitute a kind of 'praxiology', or science of successful operation, for Aristotle's 'science-as-right-talking' model of natural inquiry. Indeed, according to Bacon, natural philosophy consists in 'the inquiry of Causes and the Production of Effects' (*De augmentis*, III, 4, in Bacon 1875: 346). If the first part of this definition accords with Aristotle's idea of science, the second part is wholly original and specific to Bacon's thought, 'the link between the two parts of the definition pointing to the intimate relationship between objects of cognition and objects of construction – in short to the maker's knowledge ideal' (Perez-Ramos 1988: 106). Bacon may therefore be said to take a constructivist stance with regard to scientific truth by 'positing an alternative criterion of knowledge as operation or construction' (ibid.): 'vere scire est per causas producere posse.'

To know a given nature, therefore, is one and the same thing as to be able to produce or reproduce it.[8] But this means that the man who possesses knowledge of the natural world can also alter and even supplement nature. In other words, new knowledge translates into unprecedented effects, to such an extent that one may actually become capable of creating new 'natures'. Nowhere is this refashioning of the natural world more evident, perhaps, than in *New Atlantis*, where 'the effecting of all things possible' – which is the professed aim of Salomon's House, the scientific college presiding over Bacon's fictional island of Bensalem – leads to the fashioning of a 'new' and 'better' nature, i.e. one that is more suited to the needs and desires of man. Describing the various activities and achievements of his brethren, the Father of Salomon's House makes abundant use of superlative and comparative forms, as if to signify that science enables them not just to mimic or imitate the secret workings of nature, but also to enlarge or improve on nature itself:

> And we make (by art) in the same orchards and gardens, trees and flowers to come earlier or later than their seasons, and to come up and bear more speedily than by their natural course they do. We make them

---

[8] On Bacon's constructivist conception of truth, see also McCanles 2005. Showing that Bacon broke with the theory of concept-object correspondence, McCanles contends that Bacon 'unwittingly prophesied a hermeneutic theory of experimental procedure, one which has by now in molecular and astral physics become simply business as usual' (2005: 68).

also by art greater much than their nature, and their fruit greater and sweeter and of differing taste, smell, colour and figure, from their nature. And many of them we so order, as they become of medicinal use [. . .] Wherein you may easily think, if we have such variety of plants and living creatures more than you have in Europe [. . .] the simples, drugs, and ingredients of medicines, must likewise be in so much the greater variety. (Bacon 1999: 179–80)

Bacon's idea of science, influenced as it was by what Perez-Ramos calls 'the maker's knowledge tradition', therefore entails the potentially endless 'perfecting' of nature.[9] Such a transformation is both qualitative and quantitative. Not only does Bacon claim that science should aim at producing new natural objects, he also makes it clear that it will help to enhance the qualities of existing plants or animals.[10] The question, however, is whether such a transformation of nature is – rather like Russell's tongue-in-cheek definition of work[11] – really capable of 'indefinite extension' or whether it is restricted within given limits, however distant those limits may be. In fact, Bacon does not seem to posit a definite end to the transformation of nature. In an enlightening passage from *The Advancement of Learning* (1605), Bacon compares nature to the mythical figure of Proteus, explaining that art helps to actualise nature's virtual qualities by bringing out all its hidden facets:

For like as a man's disposition is never well known till he be crossed, nor Proteus ever changed shapes till he was straitened and held fast; so the passages and variations of nature cannot appear so fully in the liberty of nature as in the trials and vexations of art. (Bacon 2000a: 65)

In other words, the scientist's role is to interfere with the 'free course' of Nature so as to force her into all the shapes she is capable of assuming. As Sophie Weeks explains, Nature has to be shifted from her 'otiose condition' so as to reveal all her potentialities. The current

---

[9] Weeks argues that Bacon departs from the Aristotelian tradition in that he does not believe in the perfectibility of nature (Weeks 2007: 119). Yet Bacon also contends that nature is made to reveal all its potentialities by being 'vexed' and 'bound'. As a result, science makes nature fuller, and therefore also more perfect in the sense of more complete.

[10] On the perfecting of nature and the difference between art and nature, or rather lack thereof, in Bacon's philosophy, see my previous study of Francis Bacon's *New Atlantis*, especially chapter 6, 'The Relief of Man's Estate' (Popelard 2010: 203–24).

[11] See Russell's *In Praise of Idleness* (Russell 2015).

world as we see it is what Bacon calls 'Nature free', which, in itself, is already prone to change because Baconian matter has a strong tendency to dissolve. Such ontological instability entails not just a plurality of possible other worlds, but rather endless possibilities which are yet to be actualised, 'the current system being not a teleological necessity but only one possible state' (Weeks 2007: 131). The role of art or science is to bring out as many facets as possible, a task which Bacon calls 'the effecting of all things possible' (Bacon 1999: 177). Much as Proteus must be 'straightened and held fast', so lazy Nature must be 'vexed' in order to actualise all her facets and potentialities. This is what Bacon calls 'Nature bound', i.e. Nature shifted from her ordinary course by art or science. Because every change, however insignificant, discovers a new facet of nature, such a process is necessarily endless.

Nature is therefore capable of in(de)finite qualitative change. By discovering the true forms of the natural world, the scientist can artificially create a potentially endless number of new natural objects – whether they be new minerals, plants or living beings – as well as new properties which did not exist so far, thus dramatically changing the face of nature. In that regard, the Baconian scientist may be said to scrupulously follow Lady Macbeth's advice. 'Nature free' must be supplemented by 'Nature bound': only by being 'vexed' and 'put to the question' will Nature become more complete.[12] This, of course, is an infinite task. What is more, the qualitative transformation of Nature also leads to its quantitative augmentation for the newly created objects and properties come as supplements to, rather than replacements for, the old natural objects and properties. In Bacon's *New Atlantis*, the new plants, living creatures or drugs the scientists think into existence coexist alongside the old, more familiar European species. The reference to drugs and medicines reveals what an important therapeutic function science is to play. For the ultimate goal of the natural philosopher is to make human life not only better and more comfortable, but also longer and possibly eternal. In the *Historia vitae et mortis* (1623), Bacon expresses his confidence that 'anything which can be repaired without destroying the original whole, is, like the vestal flame, potentially eternal' (Bacon 2007: 147). The human body, Bacon believes, is one such thing, and the *Historia vitae* abounds in examples of biblical or legendary

---

[12] On the difference between 'torturing' and 'questioning' nature, see Pesic 1999. As he concludes, 'close examination shows that Bacon did not conceive of experiment as torture. The time has come to dismiss this idol' (Pesic 1999: 94).

figures who enjoyed exceptional longevity, such as the kings of Arcadia who lived for 300 years or 'a certain Dando in Illyria who lived for 500 years with none of the ailments of age' (Bacon 2007: 213).[13] Yet such extraordinary longevity need not necessarily remain exceptional, for it lies within the power of science to lengthen the life of man by working on the human body's vital spirits – spirits which Bacon compares to 'the craftsmen or workers who do everything that happens in the body' (Bacon 2007: 147). By so doing, the scientist will reverse the course of nature:

> If a man could arrange to put into an old body spirits of the kind characteristic of a young one, it is likely that this mighty wheel might put the other, lesser wheels into reverse, and turn back the course of nature. (Bacon 2007: 245)

In that particular instance, reversing the course of nature means achieving immortality. Thus, Bacon's reform of science is not only open-ended and incomplete because it is unfinished; it also gives rise to a new idea of science aiming at the endless augmentation of Nature and the removal of all boundaries – including that of mortality – for the benefit of mankind.

'Absolute Milan': Prospero's Unlimited Science?

As is well known, a similar interest in pushing back the limits of mortality also characterises Shakespeare's romances. Thaisa's quasi-magical resuscitation at the hand of the physician Lord Cerimon in *Pericles* is a case in point, as is, of course, the coming back to life of the statue of the 'dead' queen in *The Winter's Tale*. In both cases, the natural, entropic course of nature has been reversed – or so it seems, for neither Thaisa nor Hermione was really dead in the first place. Like Ferdinand in *The Tempest*, they, too, have 'received a second life' (*The Tempest*, 5.1.198). In *Cymbeline*, the Queen boasts that she has studied botany assiduously. She has, she says, been taught by the doctor how 'to make perfumes, distil, preserve' and she now wants to proceed to 'try the forces / Of these thy compounds on such

---

[13] Bacon's treatise is not entirely devoid of fanciful theories: setting out to correct the popular misconception that lions are long-lived because they are often toothless, he explains instead that leonine toothlessness 'could be caused by their rank breath' (Bacon 2007: 179).

creatures as / We count not worth the hanging' (*Cymbeline*, 1.6.15, 21–3).

Incidentally, 'to try' – which is to be taken in the sense of 'to experiment' – also happens to be one of Bacon's scientist's favourite verbs.[14] Both the Queen's language and her intended experimental method mark her out as a potential disciple of Baconian science as practised by the members of Salomon's House in *New Atlantis*. Although *New Atlantis* was not published until 1627, that is, some fifteen years after *Cymbeline* was first performed, Bacon maintained a fairly consistent stance throughout his life – so much so, in fact, that the main tenets of his philosophy may be said to have been quite clearly, if still imperfectly, articulated in some of his earliest works, such as *The Advancement of Learning*. It should be clear by now that I am not arguing for Bacon's direct influence on, let alone intellectual filiation with, Shakespeare's *Cymbeline* – or *The Tempest*, for that matter. Rather, what I am suggesting is that, in their own specific ways, both Bacon's philosophical works and Shakespeare's dramatic output bear witness to the rise of a new conception of the physical world whereby nature came to be seen as capable of being endlessly transformed and experimented upon. True to Lear's maxim that 'nothing can come of nothing' (*King Lear*, 1.1.82), Bacon did not so much think his philosophical system into existence by wrenching it from conceptual nothingness as he pieced it together by responding to, recalibrating and often rejecting ideas that had already been circulated by other thinkers such as Patrizi or Telesio, for instance.[15]

Similarly, it has been argued that *Cymbeline* is a play in which the influence of the new science makes itself felt in a number of ways. For example, Jachimo's 'fiery orbs' speech (1.6.31–7) suggests his wonder at the sheer scale and variety of the universe – an attitude which may not be entirely unrelated to Galileo's scientific discovery that the Milky Way consisted of 'cluster after cluster of new stars, never visible before' (Pitcher 2005: lxxiv). If, as John Pitcher reminds us, 'Shakespeare [. . .] wasn't a critic of or an apologist for the new science or the new world', it is also true that he was 'an artist [whose] eyes were open to imminent change' (Pitcher 2005: lxxvi–lxxvii). Nowhere is this attention to change more evident than in *The Tempest*, where Prospero's 'potent art' (5.1.50) shares many of the characteristics

---

[14] See Bacon 1999: 179.
[15] For Bacon's connection with Telesio and Patrizi, see Giglioni et al. 2016, Pesic 1999 and Weeks 2007.

of the new science as exemplified by Bacon, including its interest in boundlessness – although, as we shall see, this notion is less enthusiastically or unequivocally embraced in Shakespeare's play than in Bacon's philosophy.

In an essay addressing the discursive contexts of *The Tempest*, Barker and Hulme contend that 'intertextuality, or con-textualization, differs most importantly from source criticism when it establishes the necessity of reading *The Tempest* alongside congruent texts, irrespective of Shakespeare's putative knowledge of them, and when it holds that such congruency will become apparent from the constitution of discursive networks to be traced independently of authorial "intentionality"' (Barker and Hulme 2002: 200). In recent years, it is the discourse of English colonialism which has attracted the most critical attention on the part of scholars and directors. Although such interest in the colonial discursive 'con-text' of *The Tempest* is wholly justified from a historical and critical perspective, I would like to shift the emphasis back to its scientific 'con-text' – one that used to feature prominently in most critical discussions of the play before the rise of post-colonial studies[16] – while also linking it to the newly flourishing field of ecocriticism.

This is not to say that the scientific 'con-text' ought to be seen as the only, or even predominant, 'con-text' of the play, for one need not posit a dichotomy between Prospero as a mage or artist and Prospero as a colonial governor – if only because science can rarely be safely disentangled from politics. What I am suggesting instead is that, in its own dramatic and literary way, the play reflects a major epistemological turn which Bacon also expresses philosophically. Indeed, the same obsession with the boundless transformation of nature is at work in both Shakespeare's *The Tempest* and Bacon's reform of natural philosophy. For *The Tempest* is very much a 'threshold play' (Pitcher 2005: lxxvii): Janus-like, it looks both backward and forward.[17] The play's preoccupation with boundlessness aligns it with the kind of natural philosophy Bacon has come to exemplify, albeit in a much more ambivalent way. In that regard, Prospero is very much a hybrid figure bearing a likeness to both the conventional magus and the more modern Baconian 'man of science'.

---

[16] For a typical example of such an interest in the scientific and magical context of *The Tempest*, see Kermode's Arden introduction (Kermode 1966).

[17] As Barbara Mowat has shown, there is much of the traditional magician, wizard and even carnival illusionist in Prospero (Mowat 1981: 302). Corfield also argues that 'Shakespeare was eclectic in his frame of magical reference' (Corfield 1985: 34).

Prospero's art shares at least two salient characteristics with Bacon's idea of science: it is operative rather than verbally oriented like Aristotle's science, and it displays a strong preoccupation with the endless transformation of nature. That Prospero is determined not only to study the natural world but also to interact with it is evident from the first scene of the play. The storm is a demonstration of Prospero's 'potent art', as Miranda immediately conjectures: 'If by your art, my dearest father, you have / Put the wild waters in this roar, allay them' (1.2.1–2). In Elizabethan and Jacobean England – especially at the popular level – it was commonly believed that some individuals could harness supernatural powers in order to effect changes in the natural world. John Aubrey tells us, for instance, that Faldo, an old woman from Mortlake in Surrey, claimed that John Dee could cause a storm to abate.[18] Commanding the elements into obedience is certainly one of the play's central concerns. Yet, unlike common sense for Descartes, such considerable meteorological power can hardly be said to be the most fairly distributed thing in the world. As the boatswain sarcastically reminds Gonzalo, not everyone possesses such a useful skill: 'You are a counsellor, if you can command these elements to silence and work the peace of the present, we will not hand a rope more' (1.1.19–22). Whereas Gonzalo proves unable to control the natural world, Prospero, on the contrary, is consistently described as a 'god of power' who masters nature thanks to the help of his faithful spirit Ariel. In fact, as Whitlock points out, 'the name Prospero signifies both material prosperity and, in its classical and Renaissance sense, good weather [. . .] Prospero, then, is God on earth, an absolute and controlling force in human and meteorological terms' (Whitlock 1999: 177).

Prospero's exile from Milan happens to coincide with his conversion from Aristotle's verbal science to Bacon's much more practical natural philosophy. In Milan, Prospero tells us, his commitment to his 'secret studies' was of an essentially theoretical and speculative nature – so much so, in fact, that, as he explains, 'those being all my study, / The government I cast upon my brother / And to my state grew stranger' (1.2.74–6).

This passage is clearly predicated on a contrast between *vita contemplativa* and *vita activa*. In contradistinction to Prospero, Antonio is described as taking a much more practical and active stand. What Prospero had once brought into being, so to speak,

---

[18] 'Old goodwife Faldo [. . .] did know Dr Dee, and told me [. . .] that [. . .] he kept a great many stills going; that he laid the storm by magic' (Aubrey 1960: 90).

Antonio 'new created', including Prospero's followers, giving them new offices and causing them to transfer their loyalty. It is the usurper, not the Duke, who stands for action and the transformation of reality here: '[Antonio] new created / the creatures that were mine, I say, or changed 'em / Or else new formed 'em' (1.2.82–4). Very different, however, is Prospero's attitude on the island. No longer dedicating himself to his 'secret studies' – the adjective 'secret' denoting both the esotericism of a science reserved for a handful of initiates and the isolation that goes with it, together with a possible hint of subversion – Prospero now turns knowledge into action, thus following Bacon's precept that 'human knowledge and power come to the same thing, for ignorance of the cause puts the effect beyond reach' (Bacon 2004: 65). Conversely, therefore, knowledge of the cause puts the effect *within* reach. On leaving the privacy of his Italian cabinet and coming to the island, Prospero embraces a more active idea of knowledge: like Alonso's body in Ariel's song, Prospero's science has suffered 'a sea-change / Into something rich and strange' (1.2.401–2). Though Prospero continues to inhabit a 'cell' – a word that still smacks of seclusion – he can now be seen to constantly interact with both the inhabitants of the island and the physical world.[19]

My aim here is not to substitute the figure of a Baconian Prospero for the more traditional idea of Prospero as the artist/playwright/practitioner of (white) magic that critics have so often put forward, but rather to insist that the way Prospero interacts with nature in *The Tempest* may also – and simultaneously – be characteristic of the epistemological shift from 'science as right talking' to Bacon's more practical idea of science – what Perez-Ramos has called 'the maker's knowledge tradition' (Perez-Ramos 1988). Seen that way, *The Tempest* may justifiably be called a 'threshold play' indeed, radiating in every direction like Prospero's 'auspicious star' (1.2.182). As Laroque points out, Prospero is Shakespeare's dramatic response to Marlowe's Doctor Faustus, the name 'Prospero' being the Italian translation of the Latin 'faustus' – an ironical name considering the fate awaiting Marlowe's scholar (Laroque 2011: 8–9).[20] And yet, the kind of magic Prospero practises is both similar to and very

---

[19] See also Ariel's comment in 5.1.17, 'Your charm so strongly works 'em', and Prospero's admission that he aims at 'work[ing] [his] end upon their senses' in 5.1.53.

[20] Prospero is also reminiscent of Greene's Friar Bacon, while showing some affinity with the historical figure of Cornelius Agrippa von Nettesheim (Laroque 2011: 20–1).

different from Faustus's 'famous art / Wherein all nature's treasury is contained' (Bevington and Rasmussen 1993: 115). For whereas Faustus's magic clearly constitutes a transgression, Prospero's art is never described as being dangerously subversive – not even in Act 5, when Prospero alludes to his necromantic powers: 'graves at my command / Have waked their sleepers, oped, and let 'em forth / By my so potent art' (5.1.48–50). According to Jonathan Bate, in this passage loosely translated from Ovid, Shakespeare meant his audience to realise that Prospero's 'art' was 'the selfsame black magic as that of Medea' (Bate 1993: 252). Yet, although Prospero does abjure what he calls his 'rough magic' (5.1.50), he also explains, in the same breath, that he has 'required / some *heavenly* music (which even now I do) / To work mine end upon their senses that / This airy charm is for' (5.1.51–4, my emphasis). If Prospero's allusion to necromancy does indeed echo Medea's black magic, the adjective 'heavenly' also resonates with the end of *Doctor Faustus* in a way that suggests that, unlike Faustus, Prospero dances a fine line between what is lawful and what is not.[21] Instead of reading the Sycorax/Prospero antagonism as a binary opposition between black and white magic, it is possible to see Prospero's art as a 'purified' version of Sycorax's 'mischiefs manifold and sorceries terrible' (1.2.265). Prospero's 'art' is also more efficient than Sycorax's, since he now proves capable of enlisting Ariel's help in order to effect changes in the natural world. If it wasn't for Ariel, Prospero would be just as impotent and powerless as he was in Milan – so much so, in fact, that Ariel becomes Prospero's Achilles heel, i.e. a source of both weakness and strength, as the aerial spirit constantly threatens his master with rebellion. Consequently, if Prospero's magic is fully operative, it is because it is less 'earthly', that is to say purer and more aerial (Ariel), than Sycorax's. In other words, greater purity translates into increased efficacy. To become fully operative, such a 'dainty' (5.1.97) spirit as Ariel requires a more subtle and delicate kind of magic than the one Sycorax used to practise. In this sense, Prospero's art clearly echoes Bacon's call for a purification of natural magic. When 'purged of vanity and superstition', the philosopher explains, natural magic is the same as operative science:

---

[21] See Marlowe's *Doctor Faustus*: 'Regard his hellish fall, / Whose fiendful fortune may exhort the wise, / Only to wonder at unlawful things, / Whose deepness doth entice such forward wits, / To practise more than heavenly power permits' (*Doctor Faustus*, Epilogue, 4–8, in Laroque 1997: 260).

And here I will make a request, that for [Natural Prudence or the Production of Effects] I may revive and reintegrate the misapplied and abused name of natural magic, which in the true sense is but natural wisdom, or natural prudence; taken according to the ancient acception, purged from vanity and superstition. (Bacon 2000a: 80)

Science must become operative, and therefore transformative, too. It should be noted that Prospero's art happens to share Bacon's preoccupation with the transformation of nature. So changeable, in fact, is Prospero's island that the Italian aristocrats find it difficult to describe their physical environment with any degree of certainty. What some see as a barren rock, others consider to be a most fertile and promising tract of land.[22] In that regard, Gonzalo speaks more truly than he realises in Act 1, scene 1, for the polymorphous – if barren – land he would like to be standing on as the ship is going down unwittingly foreshadows Prospero's many-faceted island: 'Now would I give a thousand furlongs of sea for an acre of barren ground: long heath, broom, furze, anything. The wills above be done! but I would fain die a dry death' (1.1.61–4). As in Bacon's theory of matter, in which rest is only apparent, the physical qualities of the island – what Caliban calls the 'qualities o'th' isle' (1.2.340) – can only be temporarily identified or, as Bacon puts it, 'pinned to the ground', for everything on the island seems to be in a constant state of flux. Such ontological impermanence proceeds, of course, from Prospero's 'art'. In true Baconian fashion, Prospero forces nature to assume new shapes and visages, some of which are so 'rich and strange' that they lend credit to the most improbable travel books:

Sebastian: A living drollery. Now I will believe
That there are unicorns; that in Arabia
There is one tree, the phoenix' throne, one phoenix
At this hour reigning there. (*The Tempest*, 3.3.21–4)[23]

Interestingly enough, Alonso considers that 'these are not natural events: they strengthen / From strange to stranger' (5.1.230–1). Later on in the same scene, he adds that 'this is as strange a maze as e'er men trod / And there is in this business more than nature / Was ever

---

[22] See the often-quoted exchange between Gonzalo, Antonio and Sebastian in 2.1.53–9.
[23] On the same passage, see Lestringant's chapter in the present volume, p. 150.

conduct of: some oracle / Must rectify our knowledge' (5.1.245–8). This passage can of course be read in different, not necessarily incompatible, ways, but it is worth noting that, although Alonso clearly believes that some supernatural powers are at work on the island, he also describes his physical environment in terms that recall Bacon's ambition to transform nature scientifically and effect 'more than nature / Was ever conduct of'. Indeed, whereas Alonso distinguishes between the natural phenomena of his former life and the supernatural events he has witnessed on the island, Prospero prefers to think in strictly natural terms. As he answers Alonso, Prospero tries to alleviate his fears and convince him that these 'accidents' have a 'probable', i.e. natural, cause:

> Sir, my liege,
> Do not infest your mind with beating on
> The strangeness of this business; at pick'd leisure
> Which shall be shortly, single I'll resolve you,
> Which to you shall seem probable, of every
> These happen'd accidents; till when, be cheerful
> And think of each thing well. (5.1.248–54)

This 'strange maze', as Alonso calls it, does not call for a supernatural explanation, and Alonso's 'knowledge' will indeed 'be rectified'. What Alonso sees as supernatural events, miracles which contradict the usual course of nature, may in fact be given a perfectly 'natural' explanation if one realises that nature 'free' (i.e. the usual course of nature) only encompasses some of its possible visages. For her to reveal all her potentialities, Nature must be constrained and 'put to the question' through natural magic, since, of her own accord, she will only actualise a fraction of 'all things possible'. This is precisely what Gonzalo fails to understand as he daydreams about his imaginary commonwealth:[24] 'All things in common nature should produce / Without sweat or endeavour [. . .] but nature should bring forth, / Of its own kind, all foison, all abundance, / To feed my innocent people' (2.1.165–70). On the contrary, there is a distinctly operative, almost 'sweaty' dimension to Prospero's art as it involves the endless transformation of the island's physical environment and the production of effects which, though truly wonderful, turn out to be entirely 'probable' and natural in the end. In that regard, it is

---

[24] On the idea(l) of commonwealth which 'oscillates between the humanist dream and the carnivalesque' in *The Tempest*, see Lestringant's chapter in the present volume, p. 153.

significant that Ariel and Caliban should repeatedly insist on the operative nature of Prospero's art by using such terms as 'toil' (Ariel: 'is there
more toil'; 1.2.287) and 'work' (Caliban: 'Now Prosper works upon
thee'; 2.2.82; Ariel: 'Your charm so strongly works 'em'; 5.1.17).

Scholars have sometimes objected that Prospero is more of an
artist or an illusionist than a true natural philosopher, insisting that *The
Tempest* bears a close resemblance to the court masque (Wickham
1975, Whitlock 1999). In an essay entitled 'Open Secrets', Stephen
Orgel discards the notion that Prospero is a practitioner of black or
white magic, let alone Baconian science. Instead he explains that

> [t]he model for what he practises is theatre. His resources and repertoire
> include a troupe of actors, flying machines, thunder and lightning, dis-
> appearing banquets, mysterious music, ascents and descents, a masque
> of goddesses, even a closet full of costumes, the glittering apparel that
> proves so fatally attractive to Stephano and Trinculo. (Orgel 2011: 158)

Illuminating as Orgel's essay may be, Prospero's 'art' need not necessarily be categorised in terms of an either/or dichotomy, as Marjorie
Garber points out (Garber 2005: 854). Although it is true that Prospero creates visual and musical spectacles for Miranda and Ferdinand, as well as for Alonso and his entourage, I would like to point
out that creating sense-deceiving spectacles is also part and parcel of
Bacon's natural philosophy. In *New Atlantis*, Bacon has one of his
fictional men of science explain that deceiving the senses is an important aspect of their work. He goes on to say that such 'strange' shows,
which are truly natural, should not be falsely represented as miracles
'under pain of ignominy and fines':

> We have also houses of deceits of the senses; where we represent all
> manner of feats of juggling, false apparitions, impostures, and illusions;
> and their fallacies. And surely you will easily believe that we that have
> so many things truly natural which induce admiration, could in a world
> of particulars deceive the senses, if we would disguise those things and
> labour to make them seem more miraculous. But we do hate all impos-
> tures, and lies; insomuch as we have severely forbidden it to all our
> fellows, under pain of ignominy and fines, that they do not show any
> natural work or thing, adorned or swelling; but only pure as it is, and
> without all affectation of strangeness. (Bacon 1999: 183)

In other words, while theatre provides a model for Prospero's 'art', as
Orgel contends, it does not necessarily follow that Prospero cannot also

be seen as a practitioner of natural magic, since theatre and science are far from mutually exclusive practices – at least, not in Bacon's view.

There is no question that illusion lies at the heart of Prospero's 'art'.[25] But it is sometimes difficult to draw a clear line between scientifically based legerdemain – as illustrated by Dee's 'flying scarab' – and spectacular science – as represented, for example, by Drebbel's submarine.[26] The obvious difference, of course, consists in whether the effects produced by such 'art' are 'real' or not. Prospero, after all, claims that his 'business' is more 'probable' than 'strange'.[27] Whether his art is substantial or not is a vexed issue: most paradigmatic of all, perhaps, is the opening of the play. It has often been argued that the storm is not really a storm but the illusion of a storm[28] and it is true that neither Ferdinand nor Alonso is really dead, though both of them are made to believe that the other has drowned. As Ariel tells Prospero, 'not a hair perished' (1.2.218). Yet, if Prospero's science is an illusion, it is a very substantial one – one that leads to actual and lasting changes in the physical world, or so it seems, at least. For maintaining that Prospero's storm is 'insubstantial' makes it difficult to account for the next two lines spoken by Ariel: 'On their sustaining garments not a blemish, / But fresher than before' (1.2.219–20). The notion that the storm made their garments 'fresher than before' is also confirmed by Gonzalo: 'That our garments, being, as they were, drench'd in the sea, hold, notwithstanding, their freshness and gloss, being rather new-dy'd than stain'd with salt water' (2.1.66–9). If Prospero's art is an illusion, then, it is a lasting one – one that '*holds*', to quote Gonzalo. Its consequences make themselves felt long after the storm has abated, so that, in the end, it becomes virtually indistinguishable from life itself, which Prospero famously describes as being 'rounded with a sleep' (4.1.158). In other words, the 'insubstantiality' of life entails the substantiality of illusion: illusion is no less real than life because life is no more real than illusion.

---

[25] For Kernan, 'Prospero is a magician, and like that other magician-playwright, Doctor Faustus, his art takes the form of staging illusions' (Kernan 1979: 136).

[26] On Drebbel, see Colie 1955.

[27] But even such a simple criterion for differentiating between science and illusion turns out to be highly problematic, for theatrical illusion may, of course, produce very real effects. As Rosalind famously reminds us in the Epilogue to *As You Like It*, the fantasy world of the stage may project itself into the audience's *real* lives.

[28] See, for example, Egan's comment that 'the ship by which the Italians will leave the island after the end of *The Tempest* is found whole and "in all her trim" (5.1.239) because it was not split in the first place: its apparent destruction in a storm was an illusion' (Egan 2015: 9).

What you see, then, is what you get. If illusion is coterminous with reality, it follows that Prospero's 'art', like Baconian science, assumes a corrective dimension, as in the case of the courtiers' garments which are said to be restored to their former pristine condition. Science is capable of *reproducing* nature in the sense of 'bringing [it] again into material existence' (*OED* 2a). If one accepts the view that there is more to the storm – and, therefore, to Prospero's art – than mere 'deceit of the senses' (Bacon 1999: 183), as I think is the case, then *The Tempest* may be said to mark a clear departure from the rest of the Shakespeare canon. For, as Egan points out, the vast majority of Shakespeare's plays scrupulously follow the laws of entropy, thus asserting the unidirectionality of natural phenomena:

> The word undone here indicates a vainly hoped-for reversal of something that has been done, and asserts the impossibility of it [. . .] Shakespeare overwhelmingly uses undone to mean the state of having suffered a calamity that cannot be fixed [. . .] In his use of this done/undone phrasing, then, Shakespeare shows his sense of what we now call entropy. (Egan 2015: 8–9)

Prospero argues, in the same breath, that he 'has done nothing but in care of [Miranda]' (1.2.16) and that 'there's no harm done' (1.2.15). The spectacle of the storm, he continues, has been so 'safely ordered' (1.2.29) that those who sank will be *unsunk*, so to speak. *The Tempest* contradicts the usual unidirectional pattern, as do *The Winter's Tale*, *Cymbeline* and *Pericles* in staging the coming-to-life of the statue of Hermione, the awakening of the supposedly poisoned Imogen and the two successive reunion scenes between Thaisa, Marina and Pericles, which all transcend the cycle of life and death. Of course, Imogen, Thaisa, Marina and Hermione were never 'truly undone' in the first place, as previously mentioned, but all four romances substitute a two-directional pattern for the unidirectional entropy that most previous plays tended to illustrate. *The Tempest* goes even further than the other romances by alluding to the possibility that the usual course of nature may be reversed. If Prospero can rightfully claim that 'there's no harm done', it is not just because he creates sense-deceiving illusions but also because he can undo what he has 'done in care of [his daughter]' (1.2.16). In *The Tempest*, death cuts a much less formidable figure than in most other plays, to such an extent, in fact, that the words the poetical voice speaks to Death in Donne's *Holy Sonnet X* could also have been Prospero's – that is, before he abjures his magic and returns to Milan, where 'every third thought shall be [his] grave': 'Death, be not proud, though some have called thee / Mighty and dreadful, for thou art not so; / For those whom thou thinkst thou dost overthrow / Die not, poor Death, nor yet canst thou kill me' (ll. 1–4). In his *Historia vitae et mortis*, Bacon claimed he was confident that 'the course of nature could be turned back' (Bacon 2007: 245). So, it seems, is Prospero, whose powers appear just as limitless as Bacon's scientist's.

Prospero's powers, however, are only temporarily limitless, for unlike Bacon's open reform of science, which stretches off into infinity, Prospero's science has clear temporal limits. As Corfield puts it, 'Prospero has a stop-watch placed upon his project. [His power] will, it seems, only last until 6 p.m.' (Corfield 1985: 39). The appropriate mathematical symbol for Prospero's science is a closed circle, not a line, for, as the magus finally realises, life (and science) are just as 'rounded with a sleep' (4.1.158) as his circular island is 'rounded with' the sea. While it lasts, Prospero's power is limitless. But, as Prospero repeatedly insists during the course of the play, his actions seem to respond to a sense of urgency: 'The time 'twixt six and now / Must by us both be spent most preciously' (1.2.241–2), for 'my zenith doth depend upon / A most auspicious star, whose influence / If I now court not, but omit, my fortunes / Will ever after droop' (1.2.182–5). In short, unlike Bacon's man of science, though rather like Macbeth, who also feels 'cabin'd, cribb'd, confined, bound in / To saucy doubts and fears' (*Macbeth*, 3.4.23–4), Prospero is 'a man of limits' – a seemingly almighty natural philosopher, he displays a constant preoccupation with 'limits', 'confines', 'boundaries'.[29] His is a closed space, a confined territory, whether it is his Milan study or his island cell, and it does not come as a surprise that his final words should be those of a prisoner who 'must be released from [his bands]' (Epilogue, 9):

Now my charms are all o'erthrown,
And what strength I have's mine own,
Which is most faint: now, 'tis true,
I must be here confined by you,
Or sent to Naples. Let me not,
Since I have my dukedom got
And pardon'd the deceiver, dwell
In this bare island by your spell; (Epilogue, 1–8)

---

[29] See, for example, the following lines spoken by Prospero in the course of the play: 'Spirits, which by mine art / I have from their confines call'd to enact / My present fancies' (4.1.120–2).

The opening of the island's closed circle happens to coincide with the end of Prospero's power. Similarly, spatial enfranchisement translates into mortality: Prospero is not truly 'released from his bands' because he now remembers what he should have known all along, namely that he is tethered to his finite human condition. It looks as if his deeply seated predilection for 'closeness and the bettering of [his] mind' (1.2.90) did not, in the end, agree with the infinite horizons that his science opened up. According to Corfield, Prospero chose to abjure his 'rough magic' (5.1.50) after realising that his revenge project was morally flawed and totally incompatible with his ambition to engage in white magic only (Corfield 1985: 42). Alternatively, though perhaps not contradictorily, one may also argue that Prospero was never really comfortable with his art's intrinsic boundlessness. For it is Antonio, not Prospero, whom Shakespeare depicts as being obsessed with hubristic boundlessness. Prospero's predilection for 'closeness' and confinement caused Antonio's ambition to grow to such an extent that he soon aspired to become 'Absolute Milan' (1.2.109). At the end of the play, Prospero's ambition is both more humble and more 'limited' than his brother's as he seems to have reconciled himself to the idea of his own finitude, revealing that 'every third thought shall be [his] grave' (5.1.312).

Conclusion: 'Beyond all limits of what else in the world?'

I have tried to show that *The Tempest* could be said to reflect, and sometimes even anticipate, some of the theoretical ideas and 'procedures of thought' – to use Carla Mazzio's words (Mazzio 2009: 6) – that lie at the heart of Bacon's philosophy and bind together Bacon's philosophical and Shakespeare's dramatic discourse. In so doing, I have aimed at suggesting that both Shakespeare's play and Bacon's philosophical writings reflect a shift from a verbally oriented conception of knowledge to a much more operative idea of science which I have proposed to call, borrowing Perez-Ramos's phrase, 'the maker's knowledge' approach to the natural world. No longer a purely speculative pursuit, science entails the production of 'works' and 'the effecting of all things possible' (Bacon 1999: 177). In *New Atlantis* and in *The Tempest*, Bacon's and Shakespeare's men of science display a similar preoccupation with the endless transformation of nature and the transcending of natural boundaries – including death, which is no longer conceived of as an inevitable *terminus*. Yet, by drowning his book at the end of the play, Prospero eventually rejects the boundlessness inherent in his art. In so doing, he manifests his preference for, and maybe obsession with, boundaries – whether it takes the form of his taste for 'cells' and 'confined' spaces, or his meditation on death. As Claire Preston reminds us, 'new ways of thinking rarely exhibit themselves in tidy paradigm shifts; worldviews are always messy affairs with fuzzy boundaries and inconsistently held beliefs' (Preston 2015: 1,525). In that regard, *The Tempest* may be said to be multifaceted indeed: while foreshadowing, or reflecting, the impending epistemological shift from 'closeness' to boundlessness, it also looks backward to a more comfortable world in which Nature and the nature of man were still safely enclosed within reassuringly secure limits.

Part IV

Mechanical Tropes

Chapter 10

'Vat is the clock, Jack?': Shakespeare and the Technology of Time

Sophie Chiari

'Vat is the clock, Jack?' Doctor Caius asks Rugby in *The Merry Wives of Windsor* (2.3.3). The two friends are waiting for the arrival of Parson Evans after Caius has challenged Hugh Evans to a duel planned for just after dawn. Caius's question reveals that he probably fears for his life, not being a professed duellist himself. The answer he gets (''Tis past the hour'; 2.3.4), however imprecise it may sound, is enough to allay his fears: if it is past the hour, then it is time to move on. Caius will not die – at least, not yet.

This brief dialogue gives an illustration of the implicit tension raised by the new apprehension of time at the turn of the century. Indeed, as time was becoming increasingly mechanised, the image of society as a clock appeared more and more frequently in the literature of the period. And as early modern men and women tried to build their identities, they also aimed at fitting time to their needs rather than fitting their needs to time. Doing so, they became obsessed with watches, clocks and dials.[1] Tellingly, of all of Shakespeare's characters, only Falstaff purports to free himself from time constraints by seeking refuge in a never-never land with rivers of sack. Like Prince Harry in *1 Henry IV*, we may then wonder why Falstaff, notorious for his idle life, keeps asking for the time, since his enormous paunch has so far been his only timepiece:

> What a devil hast thou to do with the time of the day? Unless hours were cups of sack, and minutes capons, and clocks the tongues of bawds, and dials the signs of leaping-houses, and the blessed sun himself a fair hot wench in flame-coloured taffeta, I see no reason why thou shouldst be so superfluous to demand the time of the day. (1.2.5–12)

---

[1] The word 'dial' comes from the Latin *dies*, meaning 'day'.

Hal sarcastically suggests that Falstaff's circular, rather than linear, time[2] is regulated by the cycle of his drunken habits, his gormandising and his lust rather than by normal bodily appetites. Had brothel rituals, rather than watches, served to measure time, then Falstaff would have had good reason to count the hours/whores according to Jaques's pun in *As You Like It* (2.7.26–7). Yet Hal does not seem very much concerned with time either. In the space of a few lines, he lumps totally different timekeeping instruments together by simultaneously alluding to mechanical clocks, some of which had a hand for minutes, and to sundials, which he compares to a flamboyant 'wench'. So, by targeting Falstaff's chaotic existence, Hal unwittingly reveals his own frivolity and unability to count the hours correctly.

This concern with timekeeping actually reveals the complicated situation of early modern England. Since the break with Rome and Elizabeth's excommunication by Pope Pius V in 1570, the Church authorities had repeatedly warned against the 'abuses' of the old religion and the corruption of human nature. As a result, life was more or less reduced to a tragic countdown. Shakespeare's contemporaries had to cope with end-of-life anxieties from their early childhood, not simply because death could then happen at any time and was not limited to the throes of old age, but above all because it was impossible for them to know who was among the elect and who would join the reprobates. England, however, inherited a Catholic vision of duration and cessation that persisted for a long period of time.[3] As early as the seventh century, the monks began to divide their monastic time into hours thanks to the ringing of bells (Le Goff 1999: 1,118). From this point onwards, everyone knew when to pray, when to work and when to eat, but time remained essentially cyclical and seasonal, and one had to wait until the twelfth century for any significant improvements. Composed around 1100 and attributed to the Benedictine monk Honorius Augustodunensis, the *Elucidarium* established a well-ordered organisation of daily routine by distinguishing man's awakening, his labour, his meal and his rest from his social life. In the next century, Robert Kilwardy wrote a text entitled *De tempore* in which he defended Aristotle's objective, realistic time, and rejected the Augustinian theory of time's subjectivity. And the fact is that, in the thirteenth century, time's periodisation became more and more precise: the notion of 'century' dates back to

---

[2] For more details on the linearity vs the circularity of time in the play, see Hunter 1978: 127.

[3] For this conception, see Le Goff 1999.

that particular period (Le Goff 1999: 1,120). All this paved the way for a conception of authority that made time an instrument of power. Fourteenth-century Europe saw the emergence of monarchs whose priority was to control and monopolise time to ensure the submission of their subjects. The newly invented mechanical clocks, supposedly regular and reliable thanks to their verge-and-folio escapement, began to outshine traditional devices such as bells, dials, clepsammias and candles, without making them totally obsolete since the lower classes still relied on these older instruments to organise their daily schedules.

The first public clock in England was built in London by Italian clockmakers and it was set on the great tower of Windsor Castle in 1353. From the fifteenth century onwards, mechanical clocks became the contradictory symbol of temperance and death, producing a conception of time's ambivalence transmitted to the early modern era. Shakespeare's contemporaries were now able to go through a physical experience of time, and they became all the more preoccupied by timekeeping devices as these hitherto cumbersome objects were progressively reduced in size to reach a more convenient and handy format. First reserved to the public space, they gradually invaded the domestic sphere. Entirely portable, they inspired the myriad mottos ('Prepare to die', 'Consider your latter end', 'I shall return, but never thou', 'Horas est orandi',[4] 'Omnia vulnerant, ultima necat',[5] etc.) that encouraged the individual to respond to his/her oncoming death.

Even though one had to wait until 1600 for the first watches made in England and until 1631 for the Clockmakers' Company to establish itself in London, many Elizabethans had already heard of the domestic watch invented towards the end of the fifteenth century, since a number of farmhouses possessed a weight-driven wall clock (Burton 1958: 115). Because it gave (rich) men the illusion of controlling time, these new instruments did not simply favour the emergence of the individual, they also enhanced the importance paid to body image and body worship, and they deceivingly gave men the possibility to seize eternity in each minute, as St Augustine would have it (Le Goff 1999: 1,121).

In the light of this very brief history of timekeeping, many questions regarding Shakespeare's plays and poems spring to mind. How, in particular, does the materiality of time transpire in the canon? How did timekeeping devices, as contradictory symbols of decay

---

[4] 'It is the hour for prayer.' See Bedford, Davis and Kelly 2007: 25.
[5] 'All give wounds, the last one kills.'

and magnificence, fashion the early modern self? And what do they tell us about the existential preoccupations of Shakespeare's contemporaries? All three issues will be addressed in this chapter in order to highlight the links between the early modern technology of time and Shakespeare's view of the human condition.

Shakespearean Timekeeping: From Dramatic Tropes to Material Reality

Shakespearean time first and foremost suits the playwright's fancy. The most famous clock in the Shakespearean canon is undoubtedly that of *Julius Caesar*. Editors systematically point to the anachronism since no mechanical clock could strike the hour before the thirteenth century. Given his overall interest in science and new technologies, Shakespeare was perfectly aware of this impossibility, and yet he makes the chiming clock a focus of Act 2, scene 1:

> Trebonius: There is no fear in him. Let him not die;
> For he will live, and laugh at this hereafter.
> *Clock strikes*
> Brutus: Peace, count the clock.
> Cassius: The clock hath stricken three.
> Trebonius: 'Tis time to part. (2.1.190–3)

The intensity of that scene, set in the dead of night, owes much to the fact that, while the clock itself probably remained unseen (even though parish clock bells were among the props owned by early modern acting companies),[6] the audience could hear its loud sound and associate its gloomy resonances with the conspirators' plot.

By tinkering so blatantly with time, the playwright does not seek to represent an authentic Rome. The Shakespearean device is in fact comparable to that of several anonymous painters from the Netherlands who, in the 1530s and 1540s, used to depict Saint Hieronymus in his study with a domestic clock in the background replacing the original hourglass painted by Joos van Cleve.[7] Just like these Dutch artists,

---

[6] Philip Henslowe makes it clear in his diary that the Lord Admiral's Men owned two parish clock bells (called 'stepells'). See Foakes 2002: 319.

[7] This hourglass is represented on the left-hand side of the picture, on the window ledge. The composition of van Cleve's painting is based on Dürer's *Saint Jerome* (1521), but the setting, more elaborate, is van Cleve's invention.

Shakespeare superimposes the time of fiction on that of reality and, in doing so, he shatters dramatic illusion to show that ancient Rome and London actually merge onstage. Through a simple anachronism, his political reflection becomes relevant to Republican Rome as well as to Elizabethan London. One should add here that, if two temporal frameworks coexist in the play, this was also the case in Shakespeare's everyday life from 1582 onwards, since the Julian calendar, which had been introduced by Julius Caesar in 46 BC, had by then been replaced by the Gregorian calendar in the Catholic countries of Europe. So, in Protestant England, people had to cope with two time and date formats. The jarring clock in *Julius Caesar* symbolically reproduces this time gap and constitutes a privileged access to the play's metaphorical dimension.

Elsewhere, Shakespeare scatters references to timekeeping without any notable anachronism – further proof, if need be, that the mention of the clock in *Julius Caesar* is purposeful – but with countless linguistic ambiguities. While a 'clock' can indifferently refer to an abstract or a concrete reality, a 'dial' evokes either a sundial or the face of a clock, and a 'glass' designates a sundial as much as a mirror. The last example provides an important source of inspiration, for if Sonnet 77 presents a glass and a dial as two distinct (but correlated) objects, this distinction collapses in Sonnet 126, which addresses the poet's vanity and his fear of ageing: 'O thou my lovely boy, who in thy power / Dost hold Time's fickle glass, his sickle, hour' (ll. 1–2). The young man here can hold either a sandglass or a mirror, both of them allowing for self-regulation and self-examination, and the poem makes sense because these two possibilities subtly overlap. Thus Shakespeare, in his poems, juggles with the instruments of time to blur conventional manners of marking time.

The use of numerous and sometimes contradictory timekeeping devices should not, however, make us forget that the passing of time can be signalled by simpler means. Upon the mention of the first cockcrow, spectators can instantly know that it is time to start the day, and the mere allusion to a 'village cock' gives the audience all the necessary information about the setting of the play, as in the last act of *Richard III*: 'The early village cock / Hath twice done salutation to the morn' (5.3.163–4). When the playwright needs to move away from rural surroundings while keeping a sense of the rhythms of simple life, he resorts to candles, whose evocative power owes much to a triple usage: that of indicating night-time on the public stage, that of providing light, and that of symbolising time's fleetingness (Kinney 2004: 82). All three uses are relevant in Macbeth's 'Out,

out, brief candle' (5.5.22). In the same vein, an agonising Clifford exclaims in *3 Henry VI*:

> Here burns my candle out – ay, here it dies,
> Which, whiles it lasted, gave King Henry light.
> O Lancaster, I fear thy overthrow
> More than my body's parting with my soul! (2.6.1–4)

Essential here is the metaphoric interaction of light and darkness suggested by the burning candle. In this dramatic *vanitas* freezing the spectator, Clifford mentions a vacillating flame in order to allude both to the light which it gave King Henry and to his own impending *eschaton*. In fact, the very disadvantages of candles (which could be put out by draughts, and whose life duration widely differed according to the wax) aptly emphasised life's evanescence. As Arthur F. Kinney puts it, '[m]easuring time by candlelight in Shakespeare is always apocalyptic' (Kinney 2004: 82)[8] and, indeed, Elizabethans were much impressed by Saint John's Book of Revelation, which provided them with frightening details about the Day of Doom. Because candles first serve to light a space, they help disclose details that would otherwise remain invisible. They can thus be said to be 'apocalyptic' in the etymological sense of the word 'Apocalypse' (which, in Greek, meant 'the revealing of what is hidden'). Moreover, the apocalyptic symbolism of candles is in large part due to their proximity with the combined images of fire and liturgy,[9] even though Protestantism tried to suppress rituals involving the use of candles. As early as 1539, a proclamation declared that 'neither holy bread nor holy water, candles, bows, nor ashes hallowed, or creeping and kissing the cross be the workers or works of our salvation, but only be as outward signs and tokens whereby we remember Christ and his doctrine' (Simpson 2007: 22). As a consequence, in the literary production of the 1590s, the wavering flame of the candle does not only symbolise the frailty of human life, it also suggests a possible sense of nostalgia for the 'old religion'.

Shakespeare, however, does not choose between the Catholic and the Protestant visions of life and he accommodates religious differences

---

[8] See, too, Romeo's observation at the end of his clandestine night with Juliet: 'Night's candles are burnt out, and jocund day / Stands tiptoe on the misty mountain tops' (3.5.9–10).

[9] A 1550–1 inventory of St Edmund's Parish in Salisbury mentions, for example, '[c]andles to ring seven o'clock and five o'clock, and for the masons to work by'. See Cressy and Ferrell 1996: 43.

by giving precedence to subtle contrasting effects over dogmatic contents. One of the few standard features found in almost all his plays is the attempt to desacralise time, as he generally pits the tragic aspect of fleeting time against a more comical vision of temporality owing much to popular emblems and proverbs. In *King John*, he blurs the limits between the emblematic representation of time and its actual embodiment by the sexton:

> King John: France, thou shalt rue this hour within this hour.
> Bastard: Old Time the clock-setter, that bald sexton Time,
> Is it as he will? – Well then, France shall rue. (3.1.249–51)

The sarcastic tone of the Bastard suggests that Shakespeare's contemporaries, being aware of the technological advances of mechanical watchmaking, took the sexton less and less seriously. In fact, the unreliability of sextons was so notorious that it was almost proverbial. Calvin himself is said to have declared in one of his sermons: 'God makes the dayes; but Martin (the Clock-keeper) makes the howers, as he pleases' (Younge 1646: 99, quoted in Stern 2015: 15).[10]

If liturgical time, essentially cyclical in nature, worked for Shakespeare as an acoustic equivalent of the *trompe-l'œil* in painting, it is because another conception of temporality was gaining ground. The new, mechanical recording of time that pervaded the Elizabethan era enabled a more linear perception of the years passing by. Early modern horology symbolised the notion of progress rather than the idea of repetition, and it put forward the impending triumph of productivity and discipline – a military virtue cherished by the nobility. In *1 Henry VI*, the defeated French do not hesitate to compare the English weapons to a perfectly adjusted clock:

> René: I think by some odd gimmers or device
> Their arms are set, like clocks, still to strike on,
> Else ne'er could they hold out so as they do. (1.3.20–2)

The clock is here a metaphor for military coordination, and one may interpret these lines as a reluctant tribute paid to the English forces. Yet one should also take into account the notorious defectiveness of

---

[10] Similarly, according to John Taylor, the ringing of the pancake bell on Shrove Tuesday generally started at eleven o'clock, but with 'the helpe of a knauish Sexton' one could make sure that the bell rang 'commonly before nine' (Taylor 1630: sig. L4r).

mechanical objects towards the end of the century. Consequently, King René d'Anjou's image of temporal measurement actually raises questions as to how long the adversaries may remain so efficient. The remarks of the French, in sum, hint at an ironical form of distrust, for René is necessarily aware that mechanical clocks cannot strike forever.

This irony disappears when the playwright focuses on less sophisticated instruments of time. When he resorts to the 'glass'[11] to mark the duration of time, his tone becomes more serious and lyrical. The sand pouring through the hourglass is sufficient to give it a poetic quality and its mere contemplation fills the mind with dreamy visions of sea voyages and escapes. In *The Merchant of Venice*, Salerio muses:

> I should not see the sandy hour-glass run,
> But I should think of shallows and of flats,
> And see my wealthy Andrew, decks in sand,
> Vailing her hightop lower than her ribs
> To kiss her burial. (1.1.25–9)

The sand in the hourglass conjures up a shipwreck provoked by 'shallows' and 'flats'. Now, against all odds, sand was the most complex component of the hourglass. Apart from grains of sand, elements such as marble dust, powdered eggshells and even iron filings were used to fill the vial (Glennie and Thrift 2009: 285).

The 'sandy hour-glass' was familiar to Shakespeare's contemporaries as it was commonly sold by chapmen and small shopkeepers, and was also part of the church service. Indeed, when the preacher started his exhortation, he generally turned an hourglass, for sermons were then limited to one hour, no less (or the churchman was regarded as lazy), and no more (or he was seen as boring because he 'wearied [the] parishioners' with 'tedious homil[ies]'[12]). Needless to say, in pulpit practice as in drama, the presence of an hourglass necessarily implied a form of melancholy and introspection — an association famously emphasised by Dürer, whose 1514 *Melencolia* carries an hourglass.

---

[11] The first occurrence of 'glass' meaning 'hourglass' dates back to 1518, according to the *OED*: '*Cocke Lorelles Bote*, sig. C.j, One kepte ye compas and watched ye our glasse.'

[12] *As You Like It*, 3.2.152–3.

In *1 Henry VI*, the trickling of the sand in the hourglass corresponds both to a one-hour period and to a full lifetime:

> Talbot: For ere the glass that now begins to run
> Finish the process of his sandy hour,
> These eyes that see thee now well colourèd,
> Shall see thee withered, bloody, pale, and dead. (4.2.35–8)

Talbot's image of the hourglass is inserted in a rather long speech reminiscent of a sermon and it is particularly moving because it becomes a metonymy of death. However, the 'sandy hour' which sounds tragic in this context was in fact quite popular with Shakespeare's contemporaries. Indeed, in spite of their fragility and their ominous symbolism, hourglasses were widespread because they provided people with convenient portions of time. Unable to measure minutes and half-hours, they encouraged personalised schedules and gave people a certain form of freedom. In the classroom, the schoolmaster often resorted to an hourglass to start his lesson after making sure that all the pupils were there.[13] Early modern pedagogues thus did not have to cope with any outside interference as far as the regulation of time was concerned (Stern 2015: 4). They could decide when to start, regardless of the clock.

That the hourglass permits some accommodations with temporality can be seen in *The Winter's Tale*, in which Shakespeare presents us with characters living 'in an undifferentiated *longue durée*' (Orgel and Keilen 1999: 281). Sixteen years elapse between Hermione's pseudo-death and her unexpected rebirth at the end, a temporal hiatus that structures the play as a whole. The allegorical chorus of Time holding his hourglass ('I turn my glass, and give my scene such growing / As you had slept between'; 4.1.16–17)[14] solves the problem of unity raised by this sixteen-year gap. Indeed, what

---

[13] The same logic prevailed for the tutoring of the children of the nobility. Thanks to a 1547 inventory of Henry VIII's palaces, we know that young Prince Edward had a bone hourglass at his disposal to check the duration of his tutor's lessons (Pollnitz 2015: 150).

[14] On Time as a bald, bearded old man holding an hourglass, see Henry Peacham, 'The Second Book of Drawing', in *The Gentlemans Exercise (. . .) in Lyming, Painting, etc.* (1612). Peacham describes Time as follows: 'Hee is commonly drawne upon tombes in Gardens, and other places an olde man bald, winged with a Sith and an hower glasse.' Quoted as an appendix in Pafford 1963: 168 (*The Winter's Tale*).

happens before and after Act 4, scene 1 metaphorically constitutes 'the two halves of Time's hourglass' (Kiefer 2003: 162). Moreover, the trope of the glass insists on the God-like powers of the playwright, who can apparently dominate the sands of time for the sole purpose of giving Leontes a much-needed lesson in patience and temperance.

If there was a new and widespread desire for more precision at the turn of the century, Jacobean timekeeping was nonetheless of limited accuracy. We have already seen that the measuring of time through an hourglass was rather inaccurate and imprecise. As virtually anyone could manipulate its vials, a supposed sixty minutes often amounted to much more than one hour. This temporal inaccuracy characterises *The Tempest* (from the Latin *tempestas*, meaning 'weather', and *tempus*, 'time') which, contrary to *The Winter's Tale*, insists on the brevity of time.[15] In its first act, Prospero and Ariel define the temporal frame of events as follows:

> Prospero: What is the time o'th'day?
> Ariel: Past the mid season.
> Prospero: At least by two glasses. The time 'twixt six and now
> Must by us be spent most preciously. (1.2.239–41)

Prospero clearly envisions here a four-hour span (Dutton 2016: 74) and relies on the nautical method that was traditionally used to measure time. Yet, if *The Tempest* apparently sticks to the Aristotelian rule of the 'three unities' (of time, place and action), the numerous time references scattered throughout its scenes prove quite contradictory. In the last act, for example, the boatswain says that 'our ship, / Which but three glasses since we gave out split, / Is tight and yare' (5.1.225–7), which does not corroborate Prospero's initial statement and suggests instead a three-hour span – the supreme irony being that *The Tempest* is one of the few Shakespearean plays which could actually be performed within only two hours (Dutton 2016: 74). One can therefore deduce that Shakespeare chose the hourglass as the main instrument to measure the play's time not just for its escapist suggestions, but also for its notorious capacity to dilate or compress the hours along with a blessed absence of scientific precision.

---

[15] In his 2011 *Tempest* at the Theatre Royal, Haymarket, Trevor Nunn decided to place a large hourglass on the stage so that the audience could contemplate the sands of time ineluctably running.

When the playwright needs to give us a much more accurate sense of time to dramatise its painful brevity, he knows how to proceed and resorts to 'minutes', which in his works crops up more than sixty times to designate the sixtieth part of an hour (Cohen 2012: 715).[16] 'Now, at the latest minute of the hour, / Grant us your loves' (5.2.778–9), Ferdinand begs the French maids in *Love's Labour's Lost*, as the ladies are about to depart. The ultimatum voiced by the King of Navarre corresponds to the very last minutes of a comedy that, as it turns out, verges on tragedy, as no marriage has been celebrated at the end. Ferdinand's 'latest minute', at that point in the plot, suggests that the young man has acquired a keen awareness of time, something which seems to have characterised the nobility. Indeed, few people in Shakespeare's era owned the material technology to think in terms of minutes, as these only appeared on a handful of luxury clocks.[17] Needless to say, then, the word 'second' never appears in the Shakespearean canon for the simple reason that early modern horology was not yet able to fraction minutes into seconds.[18]

Shakespeare's Sounds, Sights and Bodies: From Material Reality to Dramatic Tropes

Public time became increasingly fractioned and increasingly private in sixteenth-century England. If only a small number of books devoted to measuring instruments circulated in England before the 1570s (Harkness 2007: 97–141), more and more of them were published in the late sixteenth century, thus showing that the early modern era was characterised by a rising awareness of time as a cultural concept. Learned Elizabethans now came to realise that, depending on where you lived, days were reckoned in different ways.

---

[16] The word 'minute' designating the 'sixtieth part of an hour' (*OED* 6a) dates back to the fourteenth century. The first occurrence quoted by the *OED* is drawn from John Gower's *Confessio amantis*: 'For the lachesse Of half a Minut of an houre. He loste all that he hadde do.'

[17] The fact that Shakespeare did think in terms of minutes shows his familiarity with the coterie circles of his time. Intriguingly, a few poems are numbered in the 1609 edition of his *Sonnets*, and among those, Sonnet 60 is devoted to the passage of minutes ('So do our minutes hasten to their end'; l. 2). Its being numbered seems to increase the gloomy passage of time. Such correspondence between the numbering of the sonnet and its imagery is found again in Sonnet 12, for instance, which opens on 'the clock that tells the time' (l. 1). See Graziani 1984.

[18] According to the *OED*, the word 'second' first appeared in 1589.

In a 1594 book embracing arithmetic, cosmology and navigation, Thomas Blundeville observed that

> [i]n 24. houres, which space contayneth both day and night, according to which reuolution and number of houres, the most part of Horologies or clockes in the East countryes doe goe, and are set to shew the houres of the day, but yet diuersly, for some begin their naturall day at the rising of the sunne, as the *Bohemians*, and the *Persians*, and some at the going downe of the sunne, as the *Italians*, the *Athenians*, the *Iewes*, the *Egyptians*, and the *Arabians*, but the Astronomers reckon their naturall day from noonetide to noonetide. (Blundeville 1594: 172–3)

Now, if time differs according to one's culture and religion, it also varies according to one's own subjectivity. The apprehension of time in Shakespeare is not only intimately connected to the technological discoveries of the era, it also takes into account a number of variants including personal feelings, tastes and social habits.

In effect, the playwright's representation of time is associated with a number of timekeeping devices whose material characteristics impinge on the texts. The word that most frequently recurs in the canon to designate the counting of hours is, of course, that of 'clock', whose versatility makes it suitable in different contexts. This term is primarily derived from the Latin *cloc(c)a*, meaning 'bell'. In Shakespeare's drama, the sound of ringing bells generally reminds the characters of the tyranny of time. He was also very much aware of the fact that, because bells had long been part of funeral rites, this type of acoustic memory remained vivid in spite of the changes brought about by the Reformation. In the Shakespearean canon, the ringing of bells is therefore part of a traditional dramatic economy, as in *Macbeth*, where spectators may anticipate fateful events when they hear the title character tell a servant: 'Go bid thy mistress, when my drink is ready, / She strike upon the bell' (2.1.31–2). Here, the bell is not a simple noise from the outside world, it also betrays Macbeth's tormented interior soundscape. This 'sound without a visible source' (Smith 2013: 188) can be interpreted either as a sonorous hallucination or as the perverted music of hell.

Bells were still numerous in Shakespeare's England and bellmen were a familiar sight in the streets. In London, timekeeping continued to be associated with sounds that surprised foreign visitors, especially Protestant ones who must have expected more restraint. Frederic Gerschow, who was then the secretary of Philip Julius, Duke of Stettin-Pomerania, testifies to the astonishment of the travellers who sojourned in the capital. Reporting on a visit dated 12 September 1602, he wrote:

> On arriving in London we heard a great ringing of bells in almost all the churches going on very late in the evening, also on the following days until 7 or 8 o'clock in the evening. We were informed that the young people do that for the sake of exercise and amusement, and sometimes they lay considerable sums of money as a wager, who will pull a bell the longest or ring it in the most approved fashion. Parishes spend much money in harmoniously-sounding bells, that one being preferred which has the best bells. (Gerschow 1892: 7, quoted in Smith 1999: 53)

Yet, even though other travellers like Paul Hentzner corroborate Gerschow's description, their testimonies remain biased and only represent a declining aspect of the London soundscape at the turn of the century. Indeed, the ringing of church bells gradually decreased under the influence of the Reformation, and more often than not bells only served to indicate the beginning of the service. While their ringing signalled the presence of grave illnesses in the parish, they were no longer tolled to announce the death of a parishioner (Smith 1999: 53).[19] The Puritans, whose influence was increasing, sought to privatise death rituals and silence those bells previously considered as the 'sounds of God's warnings' (MacKinnon 2015: 84).

Shakespearean drama thus retains the memory of a vanishing acoustic world. Yet, as examined above, it also takes stock of the technological advances of the age. The new mechanisms devoted to timekeeping, for Protestant ears, had the advantage of being almost silent. So, paradoxically, whereas the English Church professed a theoretical disbelief in images, it actually tipped the scales in favour of sight at the expense of hearing. As a result, the melodious and celestial roundness of tolling bells tended to be displaced by the circularity of mechanical clocks and watches – a circularity which perpetuated the ancient cosmic symbolism of the turning wheel that served to register the passage of time (Sawday 2007: 138).

---

[19] Yet the memory of it must have been kept. 'For whom the bell tolls', a phrase made famous by Ernest Hemingway, is actually lifted from one of John Donne's meditations. In a later sermon, Donne asked: 'is there any man, that in his chamber hears a bell toll for another man, and does not kneel down to pray for that dying man?' Quoted in Edwards 2001: 172.

At the bottom of the social scale, the poor only had visual access to time thanks to the large public sundials on the walls of their parish church and, as a consequence, they were not used to check the time of day. Conversely, those in the upper echelons of society had direct contact with time which allowed them to impose their personal schedule upon the members of the lower classes. Portable timepieces date back to the end of the fifteenth century (Ben-Menahem 2009: 1,358). Being relatively heavy, the first watches were hung from a girdle, or belt, and by Shakespeare's time they could finally be carried in the pocket. Their escapement mechanism was still imperfect as it depended upon 'a crown wheel and pallets with a balance ending in small weights' (ibid.). Yet owning a compact watch in early modern England was a privilege that reconciled the refined pleasure of the aesthete with the anxiety generated by the contemplation of finitude. Wearing a watch was the equivalent of a permanent memento mori, the preserve of the more affluent who could then think of their last end and work to the salvation of their souls.

In keeping with this, portable watches were the privilege of the wealthy urban spheres since a horologe cost an average twenty pounds of the time (something in the realm of £4,000 today).[20] Among the wealthier Elizabethan subjects, those who travelled often wore a small ring dial.[21] While lacking precision, this artefact was conveniently worn on the finger and one simply had to adjust the relevant date on the ring and observe where the ray of the sun, projected through a pinhole, hit the set of hour lines on the inner surface of the ring. Such devices were nevertheless heavily dependent on the weather and, of course, could be used neither inside nor at night. For all their imperfections, however, these new portable objects were much valued and served as gifts in aristocratic circles (Feingold 1984: 197). Possessing these tiny and finely engraved artefacts was in fact a means of displaying one's social rank. In *As You Like It*, Jaques seems particularly impressed by his first meeting with Touchstone because the latter sports a pocket dial indicative of his belonging to the higher spheres:[22]

---

[20] On the cost of watches, see Garber 2002: 249. According to Ian Mortimer, in the 1580s the cost of a simple watch amounted to roughly five pounds, but this seems underestimated (Mortimer 2012: 142).

[21] The future Elizabeth I had asked William Buckley to make such a ring for her. She was presented with the object in 1546.

[22] If pocket watches were extremely expensive, portable dials were more affordable. Being neither poor nor rich, Touchstone could well own a small ordinary dial.

> A fool, a fool! I met a fool i' th' forest,
> A motley fool – a miserable world! –
> [. . .]
> And then he drew a dial from his poke,
> And looking on it with lack-lustre eye
> Says very wisely 'It is ten o'clock.'
> 'Thus may we see', quoth he, 'how the world wags.' (2.7.12–23)

Because there is no clock in the forest, Touchstone probably checked the time of day while making sure that he could be seen looking at his pocket dial. Jaques has not yet spoken to Touchstone at this point in the play, but this small object is for him the sign of the fool's intimacy with the court ('One that hath been a courtier'; 2.7.36).[23]

In *Twelfth Night*, the stern Malvolio imagines himself the happiest of men in his hopes of marrying his mistress, the countess Olivia, which will then allow him to climb the social ladder. As he dreams aloud, he imagines himself exhibiting his watch in front of her: 'I frown the while, and perchance wind up my watch, or play with my – some rich jewel' (2.5.59–61). The steward is probably about to utter the word 'chain' but he changes his mind because the chain he is wearing in fact points to his condition as a steward.[24] Already bitten by Sir Toby's jibes, Malvolio prefers to evoke 'some rich jewel', but the 'jewel' can also be understood as a bawdy allusion to his penis, so that it becomes here a substitute for the chain of the steward as well as for his virile member. The coveted object is conspicuously erotic, and Malvolio is not the only character in the canon to turn horology into an obscene extension of the body. The 'hour'/'whore' homophony which characterised Elizabethan pronunciation seems to have allowed bawdy jokes about time, as can be seen in *As You Like It*, where Jaques reports Touchstone's 'And so from hour to hour we ripe and ripe, / And then from hour to hour we rot and rot' (2.7.26–7). To this, we may also add Mercutio's jest in *Romeo and Juliet*: ''Tis no less, I tell ye: for the bawdy hand of the dial is now upon the prick of noon' (2.3.103–5). In this case, the clockwork refers to the 'prick', i.e. the penis, the word 'prick' first

---

[23] As to the nobility, it was used to spending considerable sums on 'gilded and jewel-encrusted clocks' (Cohen 2012: 714) which offered their owners diverse mythological vignettes engraved upon their back.

[24] At this point in the play, the actor playing the part of Malvolio can choose to stroke his chain in order to designate what he has failed to mention. See Bruster 1992: 79.

designating the small puncture which indicated noontime, or twelve o'clock, on the dial. This double entendre is immediately grasped by the Nurse, who feigns to be shocked ('Out upon you, what a man are you!'; 2.3.106).

Hal's, Malvolio's and Mercutio's bawdy innuendoes all betray an inappropriate, marginal behaviour for, at the time, women rather than men overtly displayed their dials and watches in order to talk of their luxury items in public. Well-educated men preferred to turn these miniature devices into private personal properties which contributed to eroticise their bodies, albeit unconsciously. They kept them close to the body, under their clothes, while women generally turned them into precious ornaments placed over their garments.[25]

Whether portable watches were worn on the wrist or on a chain around the neck, they were part of the new early modern practices that appropriated, mathematised and interiorised time. They became, therefore, precious attributes that articulated the contradictory desires of seeming and being. Worn around the neck, the watch lay next to the beating heart, and its owner could thus feel and compare their own irregular pulses with the more regular ticking of a mechanical timepiece (Thompson 1967: 57). This intimate contact between the mechanical artefact and one's own flesh and body paved the way for the conception of a subjective, increasingly individualised sense of time in the sixteenth and the seventeenth centuries.

No wonder, then, that in Shakespeare's England these objects were more and more described in anthropomorphic terms and images. The word 'hand' for '[a] pointer or indicator on a dial, *esp.* one on a clock or watch indicating the passing of units of time' (*OED* 10) was already in use in the mid-sixteenth century, but Shakespeare is the first to resort to the word 'prick' to designate 'the marks by which the circumference of a dial is divided' (*OED* 3b).[26] And if the word 'face', according to the *OED*, did not refer to '[t]he surface or plate which bears the marks, digits, or hands on a watch, clock, or similar dial' (*OED* 10f) before 1659, this analogy

---

[25] Just like her father and her half-brother Edward, Elizabeth I was fascinated by horology. She possessed twenty-four watches (all of them gifts) and she employed a Bavarian horologer, Bartholomew Newsam (who had already been hired by Henry VIII), for their upkeep and good functioning. On this, see Feingold 1984: 197.

[26] The first occurrence quoted by the *OED* is from Shakespeare (3 *Henry VI*, 1.4.34–5: 'Now Phaëton hath tumbled from his car, / And made an evening at the noontide prick').

was in fact already notorious in Shakespeare's days. So, the dial would then have already suggested an ageing face, as can be seen in Sonnet 104 through the poet's extended analogy between the face of the beloved and that of a clock:

> Ah! yet doth beauty, like a dial-hand,
> Steal from his figure and no pace perceiv'd;
> So your sweet hue, which methinks still doth stand,
> Hath motion and mine eye may be deceiv'd: (ll. 9–12)

Much has already been written about the ambiguity of line 10 in particular. Suffice it to say that, if the word 'figure' obviously designates the young man's face, it also suggests the image of a dial and it subtly alludes to the ciphers designating the hours on a dial.[27] As to the 'dial' mentioned in line 9, it refers to a mechanical clock which steals each passing minute and whose single hand shadows the face of the beloved, already worried about the first signs of age.

Mechanical Failures

Because horology corresponded to God's designs, it was supposed to reflect some kind of cosmic harmony and many clockmakers had this perspective in mind. The foreigners, who generally settled 'in two former ecclesiastical Liberties', namely 'the Blackfriars and St. Martin le Grand' (Harkness 2002: 149), developed clockmaking by capitalising on this odd imagery associating the mechanical and the divine essence of time. In 1598, the Flemish clockmaker Nicholas Vallin, whose family resided in the Blackfriars, managed to build an extraordinary musical chamber clock, whose melodious sounds were produced by thirteen different bells. Up to this day, this is the earliest known surviving musical clock. The idea that God was the world's great horologer is here exemplified by the perfect combination of the techniques of beating time and of measuring music (Grant 2014: passim). The harmony produced by the carillon thus enabled Shakespeare's contemporaries, if not to hear the perfection of the cosmic order, at least to listen to something like it.

If horology was modelled on the principle of the cosmic harmony of the universe, the reverse was also true. In a fairly high number of

---

[27] For a detailed commentary on these lines, see Blakemore Evans 1996: 213.

texts, the Protestant world was indeed said to mirror the mechanics of time and the earth was compared to a series of machine parts carefully assembled by God. This is what transpires in a 1587 translation by Philip Sidney and Arthur Golding, *A woorke concerning the trewnesse of the Christian religion*,[28] originally written by the French Calvinist Philippe Duplessis-Mornay:

> Surely the Skye is as the great whéele of a Clocke, which sheweth the Planets, the Signes, the howers, and the Tydes, every one in their tyme, and that which seemeth to be his chiefe wonder, proveth him to bee subject to tyme, yea and to bee the very instrument of tyme. Now, seeing he is an instrument, there is a worker that putteth him to use, a clockkéeper that ruleth him, a Mynd that was the first procurer of his moving. For every instrument, how movable so ever it be, is but a dead thing so farre foorth as it is but an instrument, if it have not life and moving from some other thing than it it selfe [. . .] O man, the same workmayster which hath set up the Clock of thy hart for halfe a score yeares, hath also set up this huge engine of the skyes for certeyne thousands of yeares. (Duplessis-Mornay 1587: 99–100)

First comes the clock, then comes the universe. In *A woorke concerning the trewnesse of the Christian religion*, the world is not only a stage, it is a large clock set by God, while each man becomes a piece of fine watchmaking. So, when Shakespeare's contemporaries had the opportunity to look at astronomical clocks, they did not so much try to see the time of day (which was in fact quite complicated to decipher) as to decode the divine design in the universe.

As sovereigns were supposed to represent God on earth, Queen Elizabeth had to display the God-like skills of a watchmaker. According to Thomas Fale, who boasted about being the author of the first English treatise on the science of time, an expert measuring of time precisely amounted to an expert governing of the commonwealth:

> Concerning the Profite of this Art, daily experience teacheth how needful it is in a well ordered Common-wealth, seeing nothing can be done in due and convenient season, where this Science is neglected: the division of the day into certaine parts or houres, (which this Arth teacheth) doth limit and allot to each action his due time. (Fale 1593: A2v–A3r)

---

[28] Philippe Duplessis-Mornay's 1581 religious treatise was entitled *De la vérité de la religion chrestienne*.

Such observations focusing on time and order are corroborated by many literary and dramatic texts of the period, as in John Webster's *The White Devil* (1612), where the servant Cornelia associates horology with the exemplary moral rectitude that early modern subjects could expect from their princes:

> The lives of princes should like dials move,
> Whose regular example is so strong,
> They make the times by them go right or wrong. (1.2.279–81, in Webster 1996: 15)

Protestant writers often associated dials and clocks with authority, regularity and order, all the more so as timekeeping instruments were supposed to provide an apt illustration of the humanist maxim *Festina lente* ('Adage 1001' in Erasmus 1964: 171–90).[29]

However, if horology fascinated the Elizabethans, it also started to raise doubts as to its reliability. One of the first problems posed by mechanical clocks was that they had to be wound on a daily basis, periodically cleaned, and frequently oiled – a routine which was sometimes neglected. A passage in *The Tempest* should be mentioned here as Shakespeare alludes to the setting of clocks in a derisive manner. As Antonio and Sebastian laugh at Gonzalo's pompous observations, a sardonic Sebastian tells his friend: 'Look, he's winding up the watch of his wit; by and by it will strike' (2.1.13–14). Rusty as his mental mechanisms may be, the old counsellor must constantly be wound forward in order to be able to give an answer, and when he strikes, it is to profess irrelevant remarks. Sebastian hammers home the point by imitating the sound of a striking clock as soon as Gonzalo opens his mouth:

> Gonzalo: Sir, –
> Sebastian: One. Tell. (2.1.15–16)

'One' reproduces the sound of a watch striking the time, and the joke suggests that, even though Antonio, Sebastian and Gonzalo are now

---

[29] We have already seen that Protestant foreigners clearly 'dominated the clock- and watchmaking industries' in Elizabethan London (Harkness 2002: 148). Calvin himself, in spite of his rejection of jewels and ornaments, had contributed to strengthening the links between Protestantism and clockmaking in Europe by allowing jewellers to remain in Geneva and to practise their art, provided they accepted making timepieces to measure the hours (Cohen 2009: 41).

lost on an unknown island, they keep thinking in courtly ways and habits.[30] More generally, it also indicates that, if the old chiming bells made the voice of God audible to human beings, the new timepieces lacked the mystery required to be taken as genuinely divine. The new English clocks used as tools for discipline were in fact desperately human, imitating man's weaknesses and failures rather than reproducing God's intents. They never struck at the same time.[31] To make things worse, they needed to be reset quite regularly and therefore demanded almost permanent attention. Even once wound forwards or backwards, they were not totally reliable.

As a matter of fact, before Christiaan Huygens's 1656 invention of the pendulum clock, hours remained imprecise: the clocks' hands moved either too quickly or too slowly under the effects of oiling, rust and dust. Each clock ran at its own rhythm, and this destroyed the idea of perfection, harmony and unity. One can therefore understand why, in *Love's Labour's Lost*, Berowne complains about women who, in his view, turn out to be as unreliable as German clocks ('A woman, that is like a German clock, / Still a-repairing, ever out of frame, / And never going aright'; 3.1.185–7). If Berowne's remark betrays his misogyny, he is nonetheless right in suggesting that an early modern watch needed to be constantly regulated and intensively scrutinised in order to work properly (Wilson 2014: 140).[32] Moreover, his irony shows that once

---

[30] Contrary to what is often supposed, the watch imitated here cannot be a repeater since the repeating mechanism was to be 'pioneered by Edward Barlow and Daniel Quare before 1700' (Scott 2015: 155), but it certainly calls to mind those in iron and copper made by Nicholas Vallin, which were 'brought into action by pulling a cord' and could already strike the quarter and the hour (Symonds 1947: 46).

[31] Ironically, in 1579, Burghley wrote to the Queen with this reality in mind in order to encourage her to marry and bear a child while she was still fertile: 'If your majesty tarry till all clocks strike and agree of one hour, or tarry till all the oars row the barge, you shall never point the time and you may slip the tide that yet patiently tarryeth for you.' Quoted in Hackett 2007: 160.

[32] It is no mere coincidence that he should mention German clocks, since one of the pioneers of miniature watches was a Nuremberg locksmith, Peter Henlein, who made small spring-powered brass clocks attached to ribbons of steel. In 1548, another German, Caspar Werner, crafted the first iron clock of that kind to be precisely dated (Rhodes 2014: 285). German clockmakers were obsessed with the privatisation of time and England's taste for portable watches owed much to southern Germany's craze for tiny timepieces.

past the initial feelings of awe and admiration generated by the advent of new technological devices, irritation began to prevail among the population. For one thing, clocks could stop abruptly. This is registered in a number of poems of the period, which acknowledge this possibility of collapse and, by analogy, that of man's sudden death. The following lines, drawn from the 1600 miscellany *England's Parnassus* and attributed to Thomas Lodge, are explicit. They testify to a stoic form of acceptance which may seem surprising given that man's 'sudden end' is at the core of the poem:

> The shadow of the clocke by motion wends,
> We see it passe, yet marke not when it parts:
> So what is mans declines, and sudden ends,
> Each thing begins, continues and conuerts. (Abott 1600: sig. O2, p. 195)

This sense of resignation may partly be explained by the fact that, in the early 1600s, Thomas Lodge converted to Catholicism, a faith whose attitude towards death was probably more serene than that of most Protestants.

Be that as it may, mechanical clocks frequently underwent extensive and expensive repair.[33] Furthermore, they were still dependent on the sundials according to which they were frequently reset. On top of that, the very qualities they had when working correctly – measurability, homogeneity and repetitiveness – were increasingly regarded as limits because they turned human beings into machines by forbidding them to vary their own tempos. So, the mechanisms that had been an inexhaustible source of positive metaphors characterising the orderliness of the Protestant society gradually came to symbolise the collapse of the individual. Richard II's memorable monologue at the end of the play clearly emphasises the inhumanity of Elizabethan clockwork:

> I wasted time, and now doth time waste me;
> For now hath time made me his numbering clock:
> My thoughts are minutes; and with sighs they jar

---

[33] Juliet Dusinberre notably discusses the importance of an 'enormous circular time-piece in the outer court at Richmond'. This dial was painted and repaired (as documented by the accounts of the Office of Works) presumably before the courtly performance of *As You Like It* in February 1599. See Dusinberre 2003: 384.

Their watches on unto mine eyes, the outward watch,
Whereto my finger, like a dial's point,
Is pointing still, in cleansing them from tears.
Now sir, the sound that tells what hour it is
Are clamorous groans, which strike upon my heart,
Which is the bell: so sighs and tears and groans
Show minutes, times, and hours: but my time
Runs posting on in Bolingbroke's proud joy,
While I stand fooling here, his Jack o' the clock. (5.5.49–60)

Mechanical clocks had just been invented when the real Richard II (1367–1400) ascended the throne. Marking the relentless passage of time, they were still objects of fascination when Shakespeare composed his play. In the wake of Duplessis-Mornay's observations, his Richard II acknowledges that he has wasted his time and spoilt his life for, as a sovereign, he has failed to set his kingdom's tempo. Now that it is too late, he teaches himself a lesson of mortality by recording 'a sense of the body as a timepiece' (Bedford, Davis and Kelly 2007: 23). The problem is that, as indicated by the chiasmus in line 49, Richard no longer rules over time but is, on the contrary, engulfed by the jaws of 'cormorant devouring time' (*Love's Labour's Lost*, 1.1.4) represented by the 'numbering clock'. It has absorbed him so completely that the king's body has been reified, mechanised, and Richard is now part of a complex mechanism presiding over men's destinies.

The clock he alludes to probably reminded the early modern audience of real, complex clocks such as the one of Wells Cathedral (Somerset), which testified to the extreme mechanisation of the liturgy, or Wimborne Minster's astronomical clock (Dorset) (Bedford, Davis and Kelly 2007: 23). At Wells Cathedral, the dial constructed by Peter Lightfoot in the fourteenth century was placed in a chapel of the northern transept. Remarkably enough, '[o]n the face the changes of the Moon and other astronomical particulars [were] represented', and just above the dial there was 'a pair of knights armed for the Tournament pursuing each other with a rapid rotatory motion' (John Davies, the verger, quoted in Cox 2008: 183). These colourful quarter jacks, whose mechanical movements seemed both fascinating and absurd, bring to mind Richard's depreciated self-image as a 'Jack o' the clock' deprived of autonomy, having no other use than that of striking the quarter-hours.

The contemplation of such clocks was linked to the medieval trope of *contemptus mundi*, which insisted on the transience of the everyday world. It did not really encourage extensive considerations on the time of day, but it caused people to reflect on their insignificance in the universe. Looking at automata was like looking at one's own repetitive life and impending end. In *Richard II*, the title character acknowledges his current powerlessness by transforming himself into a bronze, or iron, quarter jack, whereas his adversary has been elevated into a God-like role, that of the great horologer. Clearly, if the early modern mechanical clock was sometimes endowed with a form of therapeutic power, helping people to stoically accept death, it could also be, as in Richard's case, a fearful machine symbolising man's decay.

Conclusion: For an Eschatological Vision of Life

The vision of the *eschaton* shaped cultural, religious and political practices. The more enterprising individuals tried to engage with the future in all possible ways (using prophecies and predictions, paying attention to new discoveries, planning for the next decades) in order to leave time contingencies behind them, while all the others coped with the present time and led more and more regulated existences. Indeed, the improvements in early modern horology increasingly compartmentalised the lives of the Elizabethan subjects. Shakespeare's contemporaries therefore had to learn to live with the pressure of linear time, in a shadowed and saddened present. Unrelentingly, '[t]he clock upbraid[ed] [them] with the waste of time' (*Twelfth Night*, 3.1.129).

Because the beatings of sixteenth-century timepieces echoed and reproduced the heartbeats, and because they also corresponded to the rhythm of the iambic pentameter – espousing its accelerations and its unexpected stops alike – they haunted the plays and poems of Shakespeare. As mirror images of the state, the body and the heart, his clockwork devices reaffirm the emerging place of the machine in early modern society, whereas more traditional objects such as the candle or the bells stand for the remnants of England's old faith. Significantly, the technology of time is seen as essential for a correct understanding of the world and yet few, if any, of the Shakespearean characters prove able to master it. In fact, the playwright seems to suggest that, by emphasising man's loneliness before death, the constant computation of time does not lead to happiness, but contributes instead to an enduring melancholy, a 'disease' specifically attuned to sixteenth- and seventeenth-century England.[34] As a result, time instruments function in his works both as a memento mori and as an apt reminder of the *carpe diem* philosophy.

---

[34] This particular apprehension of death continued to mark the Caroline era. One of Francis Quarles's epitaphs, written in the 1630s in memory of Lady Luckyn, is particularly telling. It is 'part of the elegy Mildreiados' and it can be seen 'in the church of Abbess Roding [Essex]'. It has the graphic shape of an hourglass, and so, in the wake of Shakespeare's sonnets, it might be said to reflect the eschatological concerns of a large part of the early modern population, bound to contemplate the past as time misspent. See Höltgen 2004.

Chapter 11

'Wheels have been set in motion': Geocentrism and Relativity in Tom Stoppard's *Rosencrantz and Guildenstern Are Dead*

Liliane Campos

By placing two minor characters centre stage, *Rosencrantz and Guildenstern Are Dead* questions the possibility of a stable frame of reference. Most critical readings have interpreted this decentring as a postmodern treatment of the canon, which destabilises accepted hierarchies and relativises the dominance not only of *Hamlet*, but also of the character Hamlet within that text. The demise of heroic narrative theorised by Lyotard, the poststructuralist concept of play, and the ontological games identified as postmodern by McHale can indeed be useful to describe Stoppard's relation to his sources (Buse 2001: 51; Vanden Heuvel 2001: 223; Jernigan 2012: 20). The epistemological drift of Stoppard's dialogue seems to confirm this affiliation to postmodernism, as his Rosencrantz and Guildenstern emphasise the relativity and uncertainty of their knowledge, and the dependence of this knowledge on frames and language games. Far from being purely meta-linguistic, these considerations weave between the metaphysical and the physical, signalling Stoppard's homage to Beckett and his interest in logical reasoning and scientific knowledge.

In 1975, Clive James ascribed Tom Stoppard's taste for instability to an Einsteinian worldview in which there is 'no point of rest' (Hayman 1979: 144). This affinity with twentieth-century physics was later confirmed by plays such as *Hapgood* (1988) and *Arcadia* (1993), in which physics became a dramatic theme in its own right. Consequently, Stoppard's film adaptation of *Rosencrantz and Guildenstern Are Dead*, shot in 1990 between the critical failure of *Hapgood* and the success of *Arcadia*, has been read as confirming either the 'Newtonian real-time universe' (Sheidley 1994: 105) or the 'Non-Euclidean', 'quantum' worldview Stoppard was exploring at the time (Zaslavskii 2005: 345; Nardo 2008: 118). Yet when we examine the play, it is striking that the tropes Stoppard chooses to express this postmodern decentring and destabilising do not actually come from Einsteinian physics, but from models of instability and relativity which already appeared in *Hamlet*. Stoppard, in other words, appropriates and reactivates images of uncertainty which he finds in Shakespeare's lines, fitting them to his own twentieth-century purposes.

In his recent book *Shakespeare and the Dawn of Modern Science*, astrophysicist Peter Usher offers a cosmological reading of *Hamlet*, based on various possible references to the contemporary debate surrounding geocentrism (Usher 2010: passim). Usher points out that Rosencrantz and Guildenstern could be named after the ancestors of Danish astronomer Tycho Brahe, whose names appear in escutcheons in his portrait.[1] He suggests that they play a part in the epistemological conflict which he reads into the play, opposing Brahe's theory of geocentrism – a partially heliocentric model – and the heliocentric 'infinite universe' proposed in 1576 by English astronomer Thomas Digges, a universe no longer contained by spheres as in Tycho Brahe's and Ptolemy's models. Usher bases his allegorical reading on Shakespeare's likely acquaintance with Digges and draws on lexical considerations such as Hamlet's mention of 'infinite space'[2] and Claudius's use of the word 'retrograde',[3] which played a significant part in contemporary cosmological debates.

Paradigm shifts, dislocations of perspective and changes in frames of reference are amongst Stoppard's favourite semiotic games.

---

[1] The 1586 portrait features heraldic shields bearing the names 'Rosencrans' and 'Guildensteren'. In 1590, Tycho Brahe sent two copies of his *De mundi aetherei recentioribus phaenomenis: liber secundus* to Thomas Savile and enclosed four copies of the portrait.

[2] Hamlet contrasts this infinity with the 'prison' of Denmark, in his first witty exchange with Rosencrantz and Guildenstern: 'I could be bounded in a nutshell and count myself a king of infinite space, were it not that I have bad dreams' (2.2.256–8).

[3] Claudius uses this term to express his displeasure at Hamlet's plans to return to Wittenberg in 1.2.113–15: 'For your intent / In going back to school in Wittenberg, / It is most retrograde to our desire.'

Although Usher's allegorical reading, in which Hamlet embodies Digges's universe and Claudius the Ptolemaic system (Usher 2010: 71–2), is too systematic to be of use here, his analysis draws our attention to the cosmological debate as a context for the question of disorientation in *Hamlet* – a context which comes to life in Stoppard's rewriting. Whether or not they originally refer to cosmological questions, the various tropes through which Shakespeare expresses disorientation are activated as specifically cosmological by Stoppard, who introduces explicit references to competing heliocentric and geocentric views. His screenplay *Galileo*, written in 1970 shortly after he first attempted to turn *Rosencrantz and Guildenstern* into a screenplay (Fleming 2001: 66), explores these theories in more depth. But *Rosencrantz and Guildenstern*, first staged in 1966 at the Edinburgh Festival, already develops an empirical, experimental gaze. Stoppard's characters repeatedly attempt thought experiments in the text and later real experiments in the film. And although their reasoning often leads to syllogisms, it flirts with key areas of twentieth-century epistemology, from relativity to probability theory.

This analysis posits reframing as the defining gesture of Stoppard's successive takes on *Hamlet* and examines this device as an epistemological manoeuvre which provides the mainspring of Stoppardian comedy. The anti-heroes of *Rosencrantz and Guildenstern Are Dead* are precisely those who complain, in Shakespeare's lines, of the lack of logical frame, begging Hamlet to 'put [his] discourse into some frame' (3.2.295–6); their own death sentence, moreover, is contained in Hamlet's letter to the king of England, which *frames* them in the colloquial sense of the word. Because they cannot perceive the tragic play whose overarching structure defines them, Stoppard's characters constantly struggle to understand their own position. In the stage version, their plight is highlighted by interactions with the players, and the film adds to these frames within frames by playing with *mise-en-abyme* effects between stage and screen. Stoppard plays with both visual and conceptual frames, carefully situating the latter in time, since the experiments which appear in the film point forward to Galilean and Newtonian physics without successfully inducing them.

Rosencrantz and Guildenstern Are Dead both adapts Shakespearean tropes as metaphors for twentieth-century uncertainty and relativism, and uses sixteenth- and seventeenth-century experiments to suggest twentieth-century epistemology. Anachronism can rarely be found at the referential level of the text: it arises from these paradoxical relations, between tenor and vehicle, knowledge and experimental framework, which result in the experiments' systematic, comic failure in the film. As a result, the spectator's reading conflicts with the characters' observations, and comedy arises from the tension between what the contemporary viewer knows and what the characters suspect. Stoppard's Rosencrantz and Guildenstern, as both puzzled spectators and pseudo-scientific observers of their own fate, bring a fresh perspective to the parallel between the experimental gaze and the theatrical gaze suggested by Shakespeare's plot. And as we watch Stoppard playing his favourite semiotic game, 'it could be *this* too' (Elam 1993: 194–5), we are reminded that the outcome of any experiment, be it scientific or theatrical, is determined by the position of its observer.

Uncontrollable Wheels

The epistemological shift performed by Stoppard can be measured by the way in which he appropriates Shakespeare's images of uncontrollable movement. In both texts, the sense of helplessness befitting Rosencrantz and Guildenstern as minor characters in the tragedy is expressed through the trope of the unstoppable wheel. As a metaphor of orderly progression, the wheel's movement expresses a higher logic which can be witnessed, but not influenced, by the characters. When Polonius dispatches them to England, Shakespeare's Rosencrantz compares the smooth working of royal power to the turning of a wheel and expands on this mechanical metaphor to convey the destructive potential of a king's death:

Rosencrantz: The cease of majesty
Dies not alone, but like a gulf doth draw
What's near it with it. It is a massy wheel
Fixed on the summit of the highest mount,
To whose huge spokes ten thousand lesser things
Are mortised and adjoined, which when it falls
Each small annexment, petty consequence,
Attends the boist'rous ruin. Never alone
Did the King sigh, but with a general groan. (3.3.15–23)

Rosencrantz's speech can be read as an attempt to justify his own obedience. Although Shakespeare does not make it altogether clear whether Hamlet's friends know that they are leading him to his death, they express their allegiance to Polonius, rather than to Hamlet, in this exchange. The strong resemblance between Rosencrantz's falling wheel and the breaking wheel of 'strumpet Fortune', invoked earlier by the First Player (2.2.496), both reinforces the idea of a higher order to which he submits and lends a less positive echo to the royal authority which he seeks to protect.

Shakespeare's trope thus hovers between the perverse action of unpredictable Fortune and an image of harmony and stability destroyed – in either case, humanity must submit to its destructive action. In *Rosencrantz and Guildenstern Are Dead*, wheels are equally uncontrollable, but they have become the fixed wheels of determinism Guildenstern gloomily describes in Act 2: 'Wheels have been set in motion, and they have their own pace, to which we are . . . condemned. Each move is dictated by the previous one – that is the meaning of order' (Stoppard 1967: 28). Stoppard's spectators can easily perceive the literary wheels to which this refers, since the title of the play gives Rosencrantz and Guildenstern's deaths the status of fate, a pre-determined outcome which they can only actualise. However, the plural introduced by Stoppard makes the image suggest a clockwork universe rather than the single wheel of Elizabethan Fortune, and allows us to frame Guildenstern's misgivings in a world ruled by Newtonian physics. The metaphor is extended in Act 3, when Stoppard's characters find themselves on a ship they did not wish to board:

> Rosencrantz: I wish I was dead. (*Considers the drop*) I could jump over the side. That would put a spoke in their wheel.
> Guildenstern: Unless they're counting on it.
> Rosencrantz: I shall remain on board. That'll put a spoke in their wheel.
> (Stoppard 1967: 52)

Rosencrantz's fantasised rebellion only strengthens the sense of predetermined behaviour, as the 'spoke' he dreams up – itself a past form of the verb *speak*, which best describes the only action he can undertake – cannot defeat the clockwork, an invisible process whose measured progress is suggested here by the verb 'counting'. The circularity of his own utterances offers a pastiche of Beckett's dialogue in *Waiting for Godot* while performing the turning of the wheel of determinism.[4] This image returns when Stoppard's characters read the letter condemning Hamlet, yet choose to do nothing. Guildenstern justifies their inaction by renewed references to a circular mechanism: 'we don't know the ins and outs of the matter, there are wheels within wheels, etcetera'; 'Other wheels are turning but they are not our concern' (Stoppard 1967: 53–4, 56).

If we turn to Elizabethan cosmology, the circular movement ascribed to Fortune's wheel can be placed within a larger context of celestial spheres and of the 'epicycles' followed by planets in the Ptolemaic universe. Ptolemy's epicycles are derived from the study of the motion of wheels, and his calculations for the pathways of the planets can be described as a system of wheels upon wheels (Mazer 2010: 12). The wheel as trope may thus have cosmological connotations, and Stoppard draws on them to link Rosencrantz and Guildenstern's wheels to astronomy, using the fixed star as an image of fate:

> Guildenstern: Free to move, speak, extemporize, and yet. We have not been cut loose. Our truancy is defined by one fixed star, and our drift represents merely a slight change of angle to it: we may seize the moment, toss it around while the moments pass, a short dash here, an exploration there, but we are brought round full circle to face again the single immutable fact – that we, Rosencrantz and Guildenstern, bearing a letter from one king to another, are taking Hamlet to England. (Stoppard 1967: 49)

As Guildenstern's metaphor wavers between the behaviour of naughty schoolboys and an ontological denial of free will, it locks the characters into the circular movement of the wheel. But whereas Shakespeare's Rosencrantz placed the king at the centre of the wheel, Stoppard's Guildenstern refers to a 'fixed star'. The

---

[4] See, for instance, the progression of dialogue through variation and repetition at the beginning of Act 2: 'Estragon: All the dead voices. / Vladimir: They make a noise like wings. / Estragon: Like leaves. / Vladimir: Like sand. / Estragon: Like leaves. [. . .] / Vladimir: What do they say? / Estragon: They talk about their lives. / Vladimir: To have lived is not enough for them. / Estragon: They have to talk about it. / Vladimir: To be dead is not enough for them. / Estragon: It is not sufficient. [*Silence.*] / Vladimir: They make a noise like feathers. / Estragon: Like leaves. / Vladimir: Like ashes. / Estragon: Like leaves' (Beckett 1986: 57).

image carries echoes of lines from *Richard II*[5] and *Love's Labour's Lost*,[6] and gives the minor characters tragic status. In the Ptolemaic scheme, it is indeed impossible, as Hermione reminds us in *The Winter's Tale*, to 'unsphere' a fixed star from the sphere that holds it (Clark 1929: 121).[7] Yet the passage also discretely creates the possibility of a heliocentric model in which the fixed star might be the sun itself. In this rhetorical pirouette, postmodern decentring both negates and confirms itself, as it refuses to grant its characters any real autonomy but construes their condition as resulting from the epistemological decentring of heliocentrism. With characteristic sleight of hand, Stoppard invites us down the road of cosmological upheaval, only to bring us back 'full circle' to the immutable order of tragedy.

Shakespeare's breaking wheel of political order is thus transformed into the impossible freewheeling of free will within the determinism of a tragic/Newtonian framework. Stoppard performs a double decentring, not only by placing secondary characters at the heart of his play, but also by activating the cosmological potential of the wheel trope to shift it towards a heliocentric model in which man no longer stands at the centre of the universe. In an ironic reversal, the characters' fear now derives from the fact that the wheel is both unknowable and indestructible. The shift turns a political metaphor into an epistemological one, as fear no longer arises from the destruction of order, but from uncertainty as to which order is at work.

Ships, Compasses and the Loss of Bearings

Before Stoppard's Rosencrantz and Guildenstern find themselves literally at sea in Act 3, their disorientation within the plot is conveyed

---

[5] See the Welsh Captain's depiction of signs announcing the fall of kings in *Richard II*, 2.4.7–9: "Tis thought the King is dead. We will not stay. / The bay trees in our country are all withered, / And meteors fright the fixèd stars of heaven.'

[6] Biron mentions the fixèd star when he protests against the strict rules of the king's proposed 'academe' in the opening scene of *Love's Labour's Lost*, 1.1.88–91: 'These earthly godfathers of heaven's lights, / That give a name to every fixèd star, / Have no more profit of their shining nights / Than those that walk and wot not what they are.'

[7] See Hermione's answer to Polixenes in *The Winter's Tale*, 1.2.48–50: 'But I, / Though you would seek t'unsphere the stars with oaths, / Should yet say "Sir, no going."'

in Act 2 by the metaphor of the compass. Taking their cue from Hamlet's claim to be 'but mad north north-west' (2.2.315), they attempt to discover in which direction the wind is blowing. While Hamlet's line emphasises his unpredictability, as changeable as the wind, Stoppard's dialogue leads his characters round in syllogistic circles, until they are forced to confront the theatricality of their situation:

> Guildenstern: if we came from down there (*front*) and it is morning, the sun would be up there (*his left*), and if it is actually over there (*his right*) and it's still morning, we must have come from up there (*behind him*), and if that is southerly (*his left*) and the sun is really over there (*front*), then it's the afternoon. However, if none of these is the case –
> Rosencrantz: Why don't you go and have a look?
> Guildenstern: Empiricism?! – is that all you have to offer? [You seem to have no conception of where we stand! You won't find the answer written down for you in the bowl of a compass – I can tell you that. (*Pause*) Besides, you can never tell this far north – it's probably dark out there.]
> Rosencrantz: I merely suggest that the position of the sun, if it is out, would give you a rough idea of the time; alternatively, the clock, if it is going, would give you a rough idea of the position of the sun. I forget which you're trying to establish.
> Guildenstern: I'm trying to establish the direction of the wind.
> Rosencrantz: There isn't any wind. *Draught*, yes.
> Guildenstern: In that case, the origin. Trace it to its source and it might give us a rough idea of the way we came in – which might give us a rough idea of south, for further reference.
> Rosencrantz: It's coming up through the floor. (Stoppard 1967: 27)[8]

During this exchange, Stoppard not only brings the argument round to a comic reminder of his title characters' position on a theatre stage, he also shifts the emphasis from the relativity of a psychological state – Hamlet's sanity as dependent on the meteorological context – to the relativity of physical observation in relation to a frame of reference ('where we stand'). Rosencrantz and Guildenstern do not, indeed, know where they stand, and their attempts at observation and logical deduction are defeated by their blindness to their own status as dramatic characters.

In Act 3, Stoppard continues to play with the frame of reference by shifting Rosencrantz and Guildenstern's location *on the boards*

---

[8] In the 1967 edition of the text Stoppard indicates optional cuts between brackets.

to their journey *on board* a ship. As they sail towards England, the ship at first seems to rescue them from their Beckettian predicament. Unlike Estragon and Vladimir, they no longer need to worry about whether to leave or not: 'I'm rather fond of boats,' Guildenstern admits. 'You don't have to worry about which way to go, or whether to go at all – the question doesn't arise, because you're on a boat, aren't you?' (Stoppard 1967: 48). But this respite is short-lived and the ship rapidly becomes an image of fate: 'One is free on a boat. For a time. Relatively' (Stoppard 1967: 49). The boat embodies both the unswerving course of the tragic plot and the principle of relativity: much like Shakespeare's plot, it is an observational frame of reference which the characters cannot escape. After Hamlet's escape, Stoppard has his characters read the letter they are bearing to the king and realise that they are condemned. They receive the news of their imminent death with appalled resignation, and Guildenstern returns to his musings about the confines of the boat: 'Where we went wrong was getting on a boat. We can move, of course, change direction, rattle about, but our movement is contained within a larger one that carries us along as inexorably as the wind and current . . .' (Stoppard 1967: 59). This dependence of movement on the observational frame of reference recalls the experiment described in Galileo's 1632 *Dialogue Concerning the Two Chief World Systems*, in which Salviati points out that experiments carried out in a ship's cabin would not allow the observer to know whether or not the ship is moving, so long as its movement is uniform. The movements described are therefore relative to the frame of observation. This thought experiment allows Salviati to demonstrate our inability to perceive the movement of the earth around the sun, and Stoppard seems to be following the same logical thread when Rosencrantz concludes despondently: 'That's it, then, is it? (*No answer. He looks out front*) The sun's going down. Or the earth's coming up, as the fashionable theory has it. (*Small pause*) Not that it makes any difference' (Stoppard 1967: 61).

In both plays, Rosencrantz and Guildenstern thus fail to find the bearings that would allow them to follow Hamlet. But in Stoppard's text, the epistemological focus of the metaphor is shifted once again from a dependence on external factors (the direction of the wind) to a dependence on the observers' own motion and position (their own direction within the *plot* of the play and the *plotted* course of the ship). Just as they cannot step outside the tragic frame they depend on, Stoppard's characters cannot observe their physical frame of reference from the outside. Physical and ontological relativity meet in the metaphor of movement. When he realises Hamlet's disappearance and their lack of a goal, Guildenstern bitterly rejects the possibility of interrupting their course: 'We've travelled too far, and our momentum has taken over' (Stoppard 1967: 58). The word 'momentum', which in the sense of 'quantity of motion' only appears in English mathematics in the late seventeenth century, is one of the rare direct scientific anachronisms of the text.[9] Just as the boat's movement points forward to Galileo, Guildenstern's repeated use of the word 'momentum' (twice in Act 3) points forward to Newton's laws of motion, particularly the law of conservation of momentum (according to which the momentum of a system remains constant if there are no external forces acting upon it). But these potential anachronisms are actualised by the play's reception rather than by the text; like the proleptic title of the play, they are produced by the spectator's position in time and the knowledge it allows. They become, in other words, part of the play's dramatic irony.

Stoppard enhances this irony in his film, where Rosencrantz (Gary Oldman) *almost* makes various great discoveries from the history of physics. He almost discovers the conservation of momentum at the end of Act 1 by accidentally turning a row of hanging clay pots into a Newton's cradle – when he lifts and releases the first of the row of pots, it strikes the others, and this transmits its force to the other end and lifts the last pot up. This improvised experiment, like all the others he performs in the film, turns into a visual joke because it fails when he attempts to replicate it – in this case, breaking one of the pots when he tries to demonstrate it to Guildenstern. In the play, Guildenstern complains of Rosencrantz's 'empiricism' (Stoppard 1967: 27) as he watches him trying to establish the direction of the wind. In the film, the accusation becomes one of 'pragmatism', illustrated by a series of these pseudo-scientific sequences (*Rosencrantz* 1990: 54:28). Stoppard thus draws on the theatrical potential of experiment (Cantor 1989: passim) – a practice based on repetition and spectatorship – and uses it as a source of miniature theatrical sequences within the film. As we watch Rosencrantz *almost* discover

---

[9] According to the *OED*, the word first appears with this meaning in 1699, in John Keill's *An examination of Dr Burnet's Theory of the Earth. Together with some remarks on Mr Whiston's new Theory of the Earth.*

Newton's theory of gravitation under an apple tree, Archimedes's principle in his bathtub, or Galileo's demonstration that bodies of different mass fall at the same speed, we recognise a pattern and come to expect comic failure.

Anachronism thus becomes a visual, rather than a textual, potential in Stoppard's film. Many of the references to science, such as the wheels that 'have been set in motion' in Act 2, or the lines about the fixed star or relative movement of the sun and the earth in Act 3, are cut.[10] But their effect is preserved, either by equivalent tropes – Guildenstern's speech about wheels that have been set in motion is replaced by the Player's lines from *Hamlet* describing the breaking wheel of Fortune – or by these visual scientific jokes which replace the textual hints at future discoveries with a comic routine, underlining our privileged position as spectators who know more than Rosencrantz. While he almost discovers steam power at the beginning of Act 2, Guildenstern remains oblivious, preoccupied by their previous conversation with Hamlet. As he unwittingly pulls apart Rosencrantz's experimental set-up, he optimistically declares: 'I think we can say we made some progress.' Here, Stoppard replaces the word 'headway' in his play (Stoppard 1967: 26) with 'progress' in his film (*Rosencrantz* 1990: 50:10), and sets up an ironic parallel between the failure of scientific progress and their failure to make any progress in their observation of Hamlet. Accordingly, Rosencrantz's defeated attempts can be seen as a light side show to the main drama of incomprehension – Rosencrantz and Guildenstern's failure to grasp their own significance. The shots place them in rooms and courtyards, endless corridors, and in a cabin for most of Act 3, but never show us the whole castle or ship. This blinkered vision corresponds to their inability, unlike the spectator, to understand the whole system.

However, laying such emphasis on failure is a reductive approach to Stoppard's epistemological games, for Rosencrantz's 'trial and error' approach also works against a linear conception of scientific discovery. His glimpses of discoveries past and future suggest a more cyclical movement, in which human invention does not

---

[10] See previous quotations. Suppressed lines include Guildenstern's assertions that 'Wheels have been set in motion, and they have their own pace, to which we are . . . condemned' and that 'Our truancy is defined by one fixed star', and Rosencrantz's relativistic observation 'The sun's going down. Or the earth's coming up, as the fashionable theory has it' (Stoppard 1967: 28, 49, 61).

comply with the arrow of progress. This movement is formulated by Septimus for his pupil Thomasina in Stoppard's next play, *Arcadia*, written shortly after the screenplay *Rosencrantz and Guildenstern Are Dead*:

> Septimus: The missing plays of Sophocles will turn up piece by piece, or be written again in another language. Ancient cures for diseases will reveal themselves once more. Mathematical discoveries glimpsed and lost to view will have their time again. You do not suppose, my lady, that if all of Archimedes had been hiding in the great library of Alexandria, we would be at a loss for a corkscrew? (Stoppard 2009: 57)

Although historical coherence is respected by Rosencrantz's never quite discovering the principles he briefly suspects, his experimental approach provides a practical equivalent to Guildenstern's verbal reasoning. It reminds us that, in Stoppard's drama, 'It's wanting to know that makes us matter' (Stoppard 2009: 102): each time they fail, their bafflement not only strengthens the sense of a lacking frame of reference, it also foregrounds the epistemological drive of the script, the hermeneutic desire that turns two minor characters into the heroes of this postmodern tale.

Actors and Observers

By allowing us access to the frame of reference that he denies his characters, Stoppard draws our attention to the role of the observer in the interpretation of experiments. In her analysis of the film, Anna K. Nardo asserts that the Players 'embody' the principles of quantum physics or the mathematics of chaos theory, since they are 'feeding themselves back into their own equation' (Nardo 2008: 118). For Nardo, Rosencrantz's experiments are doomed to failure because the space he inhabits is defined by twentieth-century science rather than the classical science he almost discovers: 'Replacing meta-theatrical jokes with scientific gags, Stoppard not only has Ros and Guil miss the point of classical physics, but he has Ros intuit the wrong science to understand the space they inhabit' (Nardo 2008: 119). Once again, this interpretation emphasises failure as the main idea, but it overlooks the fact that the discrepancy in paradigms is produced by our relation to the show. As in many other plays from *Jumpers* (1972) to *Arcadia*

(1993), Stoppard produces comic enjoyment by giving the viewer an epistemic advantage.[11]

The strongest inscription of twentieth-century epistemology in both Stoppard's play and his film is not to be found in dialogue or action, but in the relation between the spectator and the theatrical experiment. The fact that the observer's interpretational frame will determine the outcome of an experiment is not a twentieth-century discovery: Hamlet's *Mousetrap* is designed to 'unkennel' Claudius's guilt (3.2.79), and the confirmation of this hypothesis, for both Horatio and Hamlet, is scripted by Hamlet's protocol, including the rhetorical framework he sets up to prepare his friend's observation. This hermeneutic framing of experiment is reproduced scientifically, in Stoppard's film, by Rosencrantz's experiments, which can be interpreted in different ways as either nonsense (by Guildenstern) or a confirmation of valid hypotheses (by the viewer). But a key aspect of the revolutions brought about by twentieth-century epistemology is the knowledge, derived from particle physics, that the act of observation itself can determine the behaviour of the observed object. The measurement of an electron's position by a microscope will modify its momentum because light affects its movement. In an article published in 1927, Werner Heisenberg gave a famous formulation of the uncertainty principle, stating that 'the more precisely the position is determined, the less precisely the momentum is known, and conversely' (Heisenberg 1983: 64). This indeterminacy is a problem of observation rather than interpretation, as observation inevitably determines reality by having an impact on the situation it tries to measure. The behaviour of light, for example, is famously characterised by wave-particle duality: although light behaves as both a particle and a wave, experimental observation can determine its behaviour into one of these two modes.

The uncertainty principle may be hovering behind Stoppard's insistence on the 'position' and 'momentum' of his characters in Act 3. By the time he shot *Rosencrantz and Guildenstern Are Dead*,

---

[11] Stoppard uses both scenic and textual devices to play with perspective: in *Jumpers*, the set is divided in two by a partition which allows the spectator to see what the main character does not see in his wife's bedroom; in *Arcadia* the plot moves back and forth between two time periods, giving the spectator access to information about the future or the past which is not available to the characters.

he had just written a play, *Hapgood* (1988), structured around the uncertainty principle and its metaphorical application to the indeterminacy of human identity. Rosencrantz's failed experiments, which work as long as they are unobserved but collapse under Guildenstern's withering glare, can be seen as comic variations on the idea of indeterminacy.

Metaphorical indeterminacy defines the relation between the spectator of the play, or viewer of the film, and the characters' ontological status. One of the most famous thought experiments in quantum physics, which exposes the problems posed by the gap in scale between particle mechanics and the rest of physics, is known as 'Schrödinger's cat'.[12] Schrödinger imagines a subatomic mechanism whose behaviour is undetermined, because a small amount of radioactive substance can trigger a radioactive-sensitive mechanism, depending on whether an atom decays or not. The atom is both decayed and undecayed until observed, but this undetermined event is a trigger that will open a flask of poison and kill a cat in a box. Until the box is opened, since the atom is both decayed and undecayed, the cat is both alive and dead (Schrödinger 1983: 157). This paradox, which can be dissolved by simply opening the box, allows Schrödinger to illustrate the problems arising when principles from particle mechanics are extended to the macroscopic level of everyday life.

The similarity between Schrödinger's and Stoppard's paradoxes is striking. The title of the play turns our knowledge of *Hamlet* into the premise of our spectatorship: Rosencrantz and Guildenstern are simultaneously dead, since *Hamlet* has already been written, and alive. The paradox arises from our spectatorship, since we bring our foreknowledge to the play. Moreover, Rosencrantz produces a syllogistic variation on the thought experiment, about halfway through both the play and the film:

> Rosencrantz: Do you ever think of yourself as actually *dead*, lying in a box with a lid on it?
> Guildenstern: No.
> Rosencrantz: Nor do I, really. – It's silly to be depressed by it. I mean one thinks of it like being *alive* in a box, one keeps forgetting to take

---

[12] Erwin Schrödinger first presented this thought experiment in an essay exploring the conceptual problems of quantum mechanics, 'Die gegenwartige Situation in der Quantenmechanik', published in *Naturwissenschaften* in 1935.

into account the fact that one is *dead* – which should make all the difference – shouldn't it? I mean, you'd never *know* you were in a box, would you? It would be just like being *asleep* in a box. Not that I'd like to sleep in a box, mind you, not without any air – you'd wake up dead, for a start, and then where would you be? In a box. That's the bit I don't like, frankly. That's why I don't think of it . . . (Stoppard 1967: 33)

As he struggles with the idea, Rosencrantz's syllogisms uncover a real paradox, summing up the ontological contradictions of their position: they both know and do not know that they are boxed in by the plot of *Hamlet*, and they are both alive and dead. Whether we construe it as an ontological disquisition, a meta-theatrical riff or a philosophical thought experiment, his speech problematises the characters' double status in the eyes of the spectator. In a typically Stoppardian twist, the paradox arises from the very act of observation.

Conclusion: Theatrical Experiments

If the metaphorical appeal of scientific discourse derives from its hermeneutic posture, its attempts to observe and to read the world, then this appeal must be particularly strong for the postmodern playwright who positions himself as a reader of past texts. The mechanical and cosmological tropes that Stoppard borrows from Shakespeare are tools with which he expresses moral and metaphysical questions, yet they also allow him to emphasise the role of the reader/spectator/observer, in theatre as in the laboratory, and the ways in which our time-specific relation to the canon determines our spectatorship.

In their meta-theatrical plots, both *Hamlet* and *Rosencrantz and Guildenstern Are Dead* investigate the spectator's influence on the signification of what they watch, but Stoppard places a fundamental uncertainty in the act of observation itself. In *Hamlet*, *The Mousetrap* leads to a recognition of guilt which can be seen as a form of *anagnorisis*; the play is an experiment, but the experiment conforms to the tragic pattern. In *Rosencrantz and Guildenstern*, however, the protagonists' scrutiny of Hamlet does not produce any answers. All they achieve is a list of 'symptoms' (Stoppard 1967: 56) that never leads to diagnosis, and they remain incapable, to the end, of recognising even themselves, incapable of telling which is Rosencrantz, which is Guildenstern. In his description of the 'ridiculous' case of the cat experiment, Schrödinger notes that it should prevent us from 'naively accepting as valid a "blurred model" for representing reality [. . .] There is a difference between a shaky or out-of-focus photograph and a snapshot of clouds and fog banks' (Schrödinger 1983: 157). In Stoppard's experiments, however, it is precisely the out-of-focus photograph, as it brings the frame in and out of view, that produces comic pleasure.

Coda: Scepticism and the Spectacular – On Shakespeare in an Age of Science

Carla Mazzio

Spectacular Science, Technology and Superstition in the Age of Shakespeare explores striking interrelationships among various networks of knowledge and belief as they inform and are informed by early modern drama. The framing of the book offers a helpful conceptual trajectory for such a large arena of investigation. By opening with explorations of theatre in relationship to astrology and demonology before turning to medicine, music and alchemy, the opening two sections ('Popular Beliefs' and 'Healing and Improving'), for example, move from large epistemological and ontological questions about causality and agency toward more particularised arenas in which humans aimed to improve the work and operations of nature.

As the final two sections ('Knowledge and (Re)Discoveries' and 'Mechanical Tropes') open up questions about both more rarified areas of knowledge such as atomism and optics and increasingly well-known forms of experimentation and technology integral to spatial and temporal measurement and representation, the volume comes to a close by raising a set of fascinating questions not simply about science and theatre, but about science and infinitude, horology and mortality, and the tropic revolutions of technological objects such as the clock, the wheel and the compass that continue to bring writers back to Shakespeare in order to think anew about the past, present and future. As such, this book traverses a rich array of topics that demonstrate how early modern forms of knowledge, technology and belief at once figure into and are figured by the works of Shakespeare and his contemporaries.

The opening two words of the volume's title, however, 'Spectacular Science', seem to me to resonate in ways that invite further reflection on the coherence of the volume and on the status of staged science in early modernity. From the very outset, Chiari and Popelard's wonderfully provocative phrase 'Spectacular Science' subtly evokes a series of possible relationships between seeing and knowing, vision and cognition, theatre and epistemology, and art and nature – whether one is considering any one (or indeed all) of the volume's key terms, 'science', 'technology' and 'superstition'. 'Spectacular', the adjectival form of 'spectacle', of course evokes at once a large-scale event generative of wonder and admiration and a distinctly theatrical, sensational and often strikingly visual show or performance. By linking science with the logic and promise of the 'spectacular', Sophie Chiari and Mickaël Popelard swiftly index a period in history when distinctly visual demonstrations of scientific effects were often inseparable from the work of theatre. From the visual dimensions of the medicalised body to the eschatological stakes of visualised time to the entwined structures of cartographic mapping and fabulous imagery, a number of essays in this volume throw the spectacular dimensions of science – on and off stage – into high relief.[1]

While conjuring a world marked by spectacular cultural transformation and new attention to marvels and operations of nature – not to mention the dazzling mechanical and technical innovations of the period – the titular twinning of science and spectacle also raises, for a twenty-first-century readership, a series of questions about the politics and status of spectacle itself. This is especially the case given longstanding philosophical, religious and cultural forms of resisting the conspicuously visual dimensions of representation. If 'the spectacular', as Pascale Drouet puts it, 'always bears the hallmark of suspicion' (Drouet 2009: xii),[2] what are we to make of 'science' conceived or executed under the sign of the stage spectacle in particular? Or to invoke the words of the famed critic of the modern 'society of the spectacle', Guy Debord, if 'the spectacle presents itself as something enormously positive, indisputable and inaccessible', and if 'the attitude which it demands in principle is passive acceptance' (Debord 1983: 12), what might *theatre* do (or have done) to spectacles of science, or scientific knowledge, power or expertise presented in potentially overpowering or inaccessible terms?

It is no secret that Shakespeare consistently mobilised the medium of drama to subject various presumptions of knowledge and belief to suspicion (as Horatio might put it, 'There needs no ghost [. . .] come from the grave [. . .] to tell us this'). But given various traditions marked by an anti-visual or anti-theatrical bias, ranging from Aristotle's resistance to distinctly dramatic spectacle to Reformation iconoclasm, it may well be that the *shadow* of the spectacular served as an engine of scepticism in Shakespearean drama. Think of the spectacles of science and the occult in Shakespeare (whether in the form of philosophers,

---

[1] I allude here specifically to essays in this volume by Lejri, Chiari and Lestringant.
[2] For a pointed analysis of spectacle, science and Shakespeare, see also Popelard 2009.

witches, ghosts, healers, fairies, magicians, or organising epistemologies related to optics, medicine, cosmology, etc.) and you are likely to encounter a dual aspect of spectacle that at once produces wonder and scepticism. Even in the relatively subtle case of *Love's Labour's Lost* in which the play's conspicuously visual dimensions are wedded to the history of optics (as Frances Yates once suggested and as Anne-Valérie Dulac substantiates anew in this volume), it is crucial to note that the comedy works to undermine relationships between seeing and knowing. It offers, that is, an anti-spectacular spectacle of sorts, rendering the science of optics subject to embodied perception itself. The meta-spectacular dimensions of Shakespearean drama may help to explain why, when it comes to matters of early modern science, as Chiari and Popelard emphasise, as is usual with Shakespeare, it is difficult to assert what the playwright's position on the subject exactly was.

Indeed, the larger issues at stake in this volume seem to me to open up questions about just what drama may have done to science (or conceptions thereof) and what science may have done to drama. To posit the meta-spectacular dimensions of Shakespearean drama as a veritable engine of scepticism may simply offer one way to postulate how science, as Chiari and Popelard put it, 'suffered a sea change' in and through its representation onstage. Conversely, however, drama may well have experienced a 'sea change' in and through its social, geographical and even structural relationship to cultures of scientific inquiry, innovation and debate in late sixteenth-century London. Shifts within communities of knowledge in early modern England could be seismic, generating a series of 'shocks' or dramas that would have found a rather hospitable home, a local habitation and a name, as it were, in the theatre – where clashing systems, thesis and antithesis, conjecture and refutation are the stock and trade of dramatic vitality. As Shakespeare 'made visible' the powers, limitations and metaphorical potential of scientific objects, ideas and practices, that is, he often exposed the *already* dramatic social and psychological stakes of early modern science.

Because early modern scientific activity could entail complex forms of modelling, demonstration, performance and hypothesising, and because science no less than drama often had to contend with pressures of finance, theology and the politics of truth, science and theatre intersected through a range of shared problems and possibilities of representation and authority. Anything but worlds apart, as this volume has suggested, scientific and theatrical enterprises shared a range of practical as well as visual or representational forms – involving aspects of collaboration, artisanal craft, experimental methods, and constant confrontations with partial knowledge and elusive truths. Indeed, the scope and depth of connections between science and drama might be suggested by the ease with which a recent description of early modern scientific culture could be transposed to describe the work of the theatre: '[d]espite the absence of a single institution to order and control it', observes Deborah Harkness, an 'urban sensibility helped London practitioners successfully investigate nature, mediate conflicts over knowledge claims, collaborate on projects, expertly adjudicate disputes over methods, train new practitioners, seek financial support from civic and court figures, and negotiate their way through the challenges of studying nature in a crowded urban setting' (Harkness 2007: 8). Of course, worlds of science and worlds of theatre had as many points of contestation as consilience, and the essays in this volume add a great deal to our understanding of the dynamic and often unexpected relationships between the two.

Historically speaking, it is certainly arguable that precisely as science was becoming more and more 'visible' in London through the late sixteenth-century surge in attention to the production, circulation, use and newly contested status of scientific objects and ideas, drama developed in 'spectacular' new ways – rendering sensational if not always specifically 'visible' the encounter between humans and the natural world and indeed between theories of how nature worked or might be known. It is fascinating to reflect upon *Romeo and Juliet*, as François Laroque does in the lead essay to this volume, as a crushing tragedy of astrological determinism at a time in England when predictive astrology was being seriously challenged if not denounced outright. For if astrology was no longer a default mode of approaching futurity, the fact that it could become a default mode of structuring the past stands as a striking object lesson in the power and mobility of paradigms even or especially at the moment of their own potential demise. Similarly, Margaret Jones-Davies's central insight in this volume – about Shakespeare's persistent investment in the idea of perfectibility even in the wake of acute scepticism about the alchemical project itself – helps us to postulate relationships between 'science' and theatre in subtle and complex ways, considering the mobility of structures of thought and aspiration from one sphere to another. Jones-Davies reflects the extent to which scientific ideas and ideals may wax even or especially as the material bases of those ideas begin to wane, and how a language of perfectibility could shift domains – from alchemy to drama to a new religion – enabling wonder while also offering a way out of an all-too-imperfect world of religious schism and rapid institutional change.

Scholars have, of course, long been attentive to relationships between theatrical power and crumbling or weakened structures of history and culture. 'The power of Shakespeare's theater', as Stephen Greenblatt puts it in *Hamlet in Purgatory*, 'is frequently linked to its appropriation of weakened or damaged institutional structures' (Greenblatt 2001: 253–4). For Greenblatt, the radical attenuation of institutional power (in his particular case, of the Catholic church and the belief in purgatory long integral to it) left a world rife with 'imaginative materials' or spectral remains 'available for theatrical appropriation' by early modern playwrights (ibid., 249). To this now well-known formulation, the essays in this volume raise new questions about the fate of 'damaged' or defunct as well as newly postulated epistemologies and knowledge practices in the wake of various strains of scepticism, Protestantism and the new science. For if the ghost of Old Hamlet signals for Greenblatt the theatrical potential of recently evacuated religious forms and beliefs, how might we read or understand the 'spectacular science' of the very same ghost: the vivid diagnosis or medical description of his own death, his startlingly corporeal account of the power of drugs or poison on the human body, his performance of a kind of autopsy as part and parcel of a call for revenge? Here we might not only reflect upon the staging of science as integral to the cultivation of epistemological scepticism, but also consider how the demonstrative and self-authorising dimensions of science were engaged with the production of affects as well as effects. Early modern medical science certainly had its own relationship with the spectacular that – under the constraints of dramatic form – could be uncomfortably self-disclosing.

'Spectacle', wrote Debord, 'is the material reconstruction of religious illusion' and its heir (Debord 1983: 20). If this is true, then science in the wake of the Reformation may well have been the heir apparent. As the reformation of religion attempted to purge spectacle from liturgical ritual and devotional practice, it may well be that scientific and technological products or effects (and not simply theatrical ones) offered a place for the eye to turn with fear and admiration as well as curiosity, wonder and a sense of the possible. At the same time, however, scientific processes of *reasoning* may have been moving in the opposite direction, away from direct sensory perception altogether. As Chiari and Popelard observe in the introduction, '[b]y the second half of the sixteenth century, more and more poets and playwrights started exploring the tricky relationships between sensory experience and scientific explanation'. By this they index, among other things, the growing scepticism about the relationship between vision and knowledge that was in fact integral to the rise of empiricism. The increasing disjunction between seeing and knowing in this context raises a further question about the possible function or functions of dramatic spectacle and sensation.

What, that is, might we make of the theatrically and rhetorically 'spectacular' dimensions of science onstage at a time when new currents of *reasoning* within astronomy, physics, meteorology and mathematics were in many ways moving further and further away from an understanding of nature accessible to the senses, the body, the human or intuitive experience of the world? In her cogent book *Losing Touch with Nature: Literature and the New Science in Sixteenth-Century England* (2014), Mary Thomas Crane in fact argues for the conspicuous movement of scientific knowledge *away* from all things sensory (not to mention spectacular) in late sixteenth- and early seventeenth-century England.[3] Crane's argument works by drawing together four well-known developments in approaches to the natural world: theories of heliocentrism, of atomism, of the unfixed cosmos (signalled by the supernova of 1572), and the increasing power of the Hindu-Arabic 'zero' in mathematics. These theories were all, of course, in varying degrees of circulation in late sixteenth- and early seventeenth-century England. Crane draws them together, however, in order to argue that early modern English literature often reflected a range of affective responses to the increasingly abstract, counter-intuitive, non-experiential models of understanding the natural world. Although just one chapter of Crane's book focuses on Shakespeare, it is worth putting Crane's overall argument into conversation with *Spectacular Science*.

Pierre Iselin's argument in this volume is that *Twelfth Night* operates in and through attention to the discourse and sensations of music (music as a kind of medicinal food, physiological force and experiential structure enabling an 'aural perspective' that complicates the logic of comedy itself). Iselin, for example, observes that 'encomiasts and censors, physicians and philosophers' all agree that 'music operates mysteriously on man's mind and affections', and, later, that '*Twelfth Night* inaugurates a form of aural perspective centred on reception – individual as well as collective reception – that heralds *The Tempest*'. Situating Iselin's argument alongside Crane's might open up further questions about the potentially *countervailing* force of theatre's multi-mediations in the face of new and in many ways inaccessible ways of thinking (mathematical and medical as well as cosmological)

---

[3] In the process of establishing the framework of such a large epistemological shift, Crane nicely complicates some all-too-familiar assumptions about the rise of modern science in early modern England. The well-known movement in early modern natural philosophy away from a text-based reliance upon ancient authorities to a new emphasis on empiricism and thus direct observation, for example, is complicated by Crane's consideration of how certain aspects of Aristotelian 'book knowledge' were in fact deeply intuitive and observationally based, and how empiricism often entailed a reliance upon forms of knowledge that exceeded the capacities of direct observation. Thus an overemphasis on the shift from ancient texts to the 'book of nature' itself can be understood to obscure a more striking movement from an intuitive, embodied approach to nature to a counter-intuitive and abstract one.

that were becoming increasingly abstract and thus inaccessible to the senses, embodied experience and intuition. Jonathan Pollock's essay on atomism itself may suggest as much. For if atomism conjured invisible materialisms beyond the reaches of the senses, Shakespeare makes spectacular, palpable and moving the very drama of 'nothing', the drama of unseen seeds of being, of infinite divisibility and emptiness, in and through the drama of *King Lear*. Rather than read *King Lear* as registering psychic or intellectual disorientation in the face of nature newly understood to operate in ways inaccessible to ordinary sensory experience, as Crane does so brilliantly, we might read it alongside Pollock's essay as a powerful act of translating atomism into a spectacular kind of sense. Although Pollock ultimately argues that 'Shakespeare uses atomist physics and ethics in order to multiply perspectives and do justice to the complexity of human experience' in his plays, what his essay also demonstrates is the extent to which the effects of invisible operations of nature, so integral to the creation and disintegration of matter, when situated onstage, come into rather spectacular 'view'. It also suggests the extent to which effects translate as affects in Shakespeare's theatres of matter, space and motion.

To now recall and slightly reformulate the opening question of Chiari and Popelard's introduction – 'what is this thing called science?' – we might ask what is this thing when it is itself a kind of 'nothing', be it an abstraction, a theory, a speculation or postulate without clear evidence? Is it not the work of the poet-dramatist, to quote Theseus in *A Midsummer Night's Dream*, to turn 'forms of things unknown' to 'shapes', and to 'give to airy nothing / A local habitation and a name' (5.1.16–17)? Essay contributions by Lestringant, Chiari and Popelard bring the volume further in the direction of the unknown as indexed onstage through representations of geographical space, mechanised time and the logic of the infinite. The unknown aspects of early modernity concerning new spaces and places and anxieties about duration and finitude often resulted in spectacular imaginings, new approaches to the shape and face of time, and possibly even new commitments to finite projects with a clear and distinct beginning, middle and end.

The Renaissance world was rooted in the foreign and the unknown, explains Lestringant, and those unknown spaces and places were mapped out in spectacular ways, with atlases depicting 'a gaping, tempestuous Hell', with maps round about with disembodied heads of old men, children and skulls emitting, respectively, breath as wind, heat as suns, or more skulls as earthly matter. Theatrical projects were also, of course, rife with projections and the unknown offered a kind of intergalactic space for speculative innovation so vividly thematised and challenged in *The Tempest*, perhaps the most conspicuously spectacular of Shakespeare's plays. Drawing on Montaigne, Lestringant suggests a rather humbling trajectory of encountering the unknown in science, itself deeply resonant with the trajectory of *The Tempest*, in which 'Admiration is the ground of all Philosophy; Inquisition the progress; Ignorance the end' (Florio 1999: 3.11).

Relationships between admiration and ignorance take a series of different turns in Chiari's approach to concepts of time in Shakespeare. As she reveals how Shakespeare consistently links the spectacle of visualised time (clocks, watches and the ekphrastic invocation of both) to the spectre of final ends, death and mortality, and the perils of a finite and yet uncertain future, she calls attention, again, to the psychic, dramatic and metaphysical dimensions of science in the form of instruments or machines. With Shakespeare, Chiari exposes the logic of the fetish of the watch or clock: the material incarnation of temporal movement might delight, liberate and fascinate while also binding the owner, wearer or observer to the inevitability of his or her own end. Conversely, Popelard's focus on the logic and rhetoric of infinity and endlessness offers a new perspective on bounded time, space and projects in Shakespearean drama – particularly marked in the figure of Prospero. These three essays combined call upon readers to meditate further on drama as a particular and particularly charged medium for expressing and managing a series of mysteries of space, time and being.

With this volume, the archive of possible methods and sources with which to understand Shakespeare's relationship to science continues to grow. Alongside refreshing approaches to texts long associated with Shakespeare, ranging from Reginald Scot's *Discoverie of Witchcraft* and Edward Jorden's *Briefe Discourse of a Disease Called the Suffocation of the Mother* to Tom Stoppard's *Rosencrantz and Guildenstern Are Dead*, the essays as a whole open up new and pressing questions about Shakespeare's relationship to popular science and discarded beliefs, to links and ruptures between religion and science, to cutting-edge developments and speculations integral to the advancement of science, and to mechanical instruments and ideas of being. Following in the models of the best of drama and the best of science, this volume asks more questions than it answers, and offers a new testing ground for a swiftly developing area within early modern and Shakespeare studies. If the meta-spectacular can produce wonder and scepticism at once, it is all the more important at this point in time that we continue to reassess the power of drama – in this case the drama of Shakespeare – to delight and move while also relieving spectacle of its power to render subjects captive, passive or isolated from the logic, politics and operations of knowledge at work in the world.

Bibliography

Abott, Robert (1600), *Englands Parnassus: or the choysest flowers of our moderne poets*, STC (2nd edn) 378.

Abraham, Lyndy (1998), *A Dictionary of Alchemical Imagery*, Cambridge: Cambridge University Press.

Abraham, Lyndy (1999), 'Weddings, Funerals and Incest: Alchemical Emblems and Shakespeare's *Pericles, Prince of Tyre*', *Journal of English and Germanic Philology*, 98.4, 523–49.

Albanese, Denise (1996), *New Science, New World*, Durham, NC: Duke University Press.

Albertus Magnus (1890), *Mineralium*, in *Opera Omnia*, ed. Augusti Borgnet, Apud Ludovicum Vives, vol. 5.

Almond, Philip C. (2004), *Daemonic Possession and Exorcism in Early Modern England: Contemporary Texts and their Cultural Contexts*, Cambridge: Cambridge University Press.

Anon. (1540), *La Pazzia*, Venice.

Anon. (1590), *Tarlton's newes out of Purgatorie. Onely such a iest as his iigge, fit for gentlemen to laugh at an houre, &c. Published by an old companion of his, Robin Goodfellow*, London: Printed [by R. Robinson] for T. G[ubbin] and T. N[ewman].

Aristotle (1985), *On Sleep*, trans. J. I. Beare, in *The Complete Works*, ed. Jonathan Barnes, vol. 1, Princeton: Princeton University Press.

Arnold, Matthew (1962) [1885], *Discourses in America*, in M. H. Abrams (ed.), *The Norton Anthology of English Literature*, vol. 2, New York: W. W. Norton.

Aromatico, Andrea (1996), *Alchimie le grand secret*, Paris: Découvertes Gallimard.

Aubrey, John (1960), *Brief Lives*, ed. Oliver Lawson Dick, London: Secker and Warburg.

Auden, W. H. (1962), 'Music in Shakespeare', in *The Dyer's Hand*, London: Random House, 500–27.

Aughterson, Kate (1998), *The English Renaissance: An Anthology of Sources and Documents*, London and New York: Routledge.

Augustine (St) (1984), *City of God*, trans. Henry Bettenson, with an introduction by John O'Meara, Harmondsworth: Penguin Classics.

Austern, Linda Phyllis (1992), *Music in English Children's Drama of the Late Renaissance*, New York: Gorgon and Breach.

Bacon, Francis (1826) [1626], *Sylva sylvarum; or, A Natural History in Ten Centuries*, in *The Works of Francis Bacon in Ten Volumes*, vol. 1, London: C. and J. Rivington.

Bacon, Francis (1875), *The Works of Francis Bacon, Vol. IV: Translation of Philosophical Works*, ed. James Spedding, Robert Leslie Ellis and Douglas Denon Heath, London: Longman.

Bacon, Francis (1999), *New Atlantis*, in Susan Bruce (ed.), *Three Early Modern Utopias*, Oxford: Oxford World's Classics.

Bacon, Francis (2000a), *The Advancement of Learning*, The Oxford Francis Bacon, vol. 4, ed. Michael Kiernan, Oxford: Clarendon Press.

Bacon, Francis (2000b), *The Essayes or Counsels, Civill and Morall*, The Oxford Francis Bacon, vol. 15, ed. Michael Kiernan, Oxford: Clarendon Press.

Bacon, Francis (2002), *The Major Works*, ed. Brian Vickers, Oxford: Oxford University Press.

Bacon, Francis (2004), *The Instauratio magna Part II: Novum organum and Associated Texts*, The Oxford Francis Bacon, vol. 11, ed. Graham Rees and Maria Wakely, Oxford: Clarendon Press.

Bacon, Francis (2007), *The Instauratio magna Part III: Historia naturalis and historia vitae*, The Oxford Francis Bacon, vol. 12, ed. Graham Rees and Maria Wakely, Oxford: Clarendon Press.

Badiou, Alain (2015), *Éloge des mathématiques*, Paris: Flammarion.

Barkan, Leonard (1981), 'Living Sculptures: Ovid, Michelangelo and *The Winter's Tale*', *English Literary History*, 48.4, 639–67.

Barker, Francis, and Peter Hulme (2002), 'Nymphs and Reapers Heavily Vanish: The Discursive Con-texts of *The Tempest*', in John Drakakis (ed.), *Alternative Shakespeares*, London: Routledge, 195–209.

Barthes, Roland (1975), *Roland Barthes par Roland Barthes*, Paris: Seuil.

Bate, Jonathan (1993), *Shakespeare and Ovid*, Oxford: Oxford University Press.

Batho, G. R. (1960), 'The Library of the "Wizard" Earl: Henry Percy Ninth Earl of Northumberland (1564–1632)', *The Library*, 5th series, 15.4, 246–61.

Beckett, Samuel (1986), *Waiting for Godot*, in *The Complete Dramatic Works*, London: Faber and Faber.

Bedford, Ronald, Lloyd Davis and Philippa Kelly (2007), *Early Modern English Lives: Autobiography and Self-Representation, 1500–1660*, Aldershot: Ashgate.

Ben-Menahem, Ari (2009), *Historical Encyclopedia of Natural and Mathematical Sciences*, Berlin: Springer-Verlag, vol. 1.

Benoit, Paul (1991), 'La théologie au XIIIe siècle: une science pas comme les autres', in Michel Serres (ed.), *Éléments d'histoire des sciences*, Paris: Bordas, 177–95.

Berger, Harry Jr (2013), *A Fury in the Words: Love and Embarrassment in Shakespeare's Venice*, New York: Fordham University Press.

Bevington, David, and Eric Rasmussen (eds) (1993), *Doctor Faustus: A- and B-texts (1604, 1616)*, Manchester: Manchester University Press.

Blakemore Evans, G. (ed.) (1996), *The Sonnets*, Cambridge: Cambridge University Press.

Blundeville, Thomas (1594), *M. Blundevile his exercises containing sixe treatises*, London, STC (2nd edn) 3,146.

Bobrick, Benson (2005), *The Fated Sky: Astrology in History*, New York: Simon & Schuster.

Boireau, Nicole (1997), 'Tom Stoppard's Metadrama: The Haunting Repetition', in Nicole Boireau (ed.), *Drama on Drama: Dimensions of Theatricality on the Contemporary British Stage*, London: Macmillan, 136–51.

Booth, Stephen (ed.) (1977), *Shakespeare's Sonnets*, New Haven: Yale University Press.

Bright, Timothy (1969) [1586, 1613], *A Treatise of Melancholy*, Amsterdam and New York: Da Capo Press/Theatrum Orbis Terrarum Ltd, The English Experience, no. 212.

Brooke, Nicholas (ed.) (1990), *Macbeth*, The Oxford Shakespeare, Oxford: Oxford University Press.

Brooks, Harold F. (ed.) (1979), *A Midsummer Night's Dream*, Arden Second Series, London: Routledge.

Brooks-Davies, Douglas (1983), *The Mercurian Monarch: Magical Politics from Spenser to Pope*, Manchester: Manchester University Press.

Bruster, Douglas (1992), *Drama and the Market in the Age of Shakespeare*, Cambridge: Cambridge University Press.

Burns, Edward (ed.) (2000), *1 Henry VI*, Arden Third Series, London: Bloomsbury.

Burrow, Colin (2016), 'Montaignian Moments: Shakespeare and the *Essays*', in Neil Kenny, Richard Scholar and Wes Williams (eds), *Montaigne in Transit: Essays in Honour of Ian Maclean*, Oxford: Legenda, 233–46.

Burton, Elizabeth (1958), *The Pageant of Elizabethan England*, New York: Scribner.

Burton, Robert (2016) [1621], *The Anatomy of Melancholy*, Oxford: Benediction Classics.

Burtt, Edwin Arthur (1925), *The Metaphysical Foundation of Modern Physical Science*, London: Kegan Paul; New York: Harcourt, Brace and Company.

Buse, Peter (2001), '*Hamlet* Games – Stoppard with Lyotard', in *Drama + Theory: Critical Approaches to Modern British Drama*, Manchester: Manchester University Press, 50–68.

Cairncross, Andrew (ed.) (1960), *The First Part of King Henry VI*, Arden Second Series, London: Methuen.

Calderwood, James (1989), *The Properties of Othello*, Amherst: University of Massachusetts Press.

Campos, Liliane (2011), *The Dialogue of Art and Science in Tom Stoppard's Arcadia*, Paris: Presses Universitaires de France.

Cantor, Geoffrey (1989), 'The Rhetoric of Experiment', in David Gooding, Trevor Pinch and Simon Schaffer (eds), *The Uses of Experiment: Studies in the Natural Sciences*, Cambridge: Cambridge University Press, 159–80.

Carey, John (ed.) (1996), *John Donne: Selected Poetry*, Oxford: Oxford University Press.

Carroll, William C. (ed.) (2009), *Love's Labour's Lost*, The New Cambridge Shakespeare, Cambridge: Cambridge University Press.

Case, John (1588), *Apologia musices*, Oxford: Joseph Barnes.

Céard, Jean (1977), *La Nature et les prodiges. L'insolite au XVIe siècle en France*, Geneva: Droz.

Chalmers, Alan F. (1982), *What Is This Thing Called Science? An Assessment of the Nature and Status of Science and its Methods*, St Lucia: University of Queensland Press.

Chen-Morris, Raz (2016), 'Astronomy, Astrology, Cosmology', in Bruce R. Smith (ed.), *The Cambridge Guide to the Worlds of Shakespeare, Vol. 1: Shakespeare's World, 1500–1660*, Cambridge: Cambridge University Press, 259–60.

Chomarat, Jacques (1981), *Grammaire et rhétorique chez Érasme*, Paris: Les Belles Lettres, vol. 2.

Clark, Cumberland (1929), *Shakespeare and Science*, New York: Haskell.

Clark, Sandra, and Pamela Mason (eds) (2015), *Macbeth*, Arden Third Series, London: Bloomsbury.

Clark, Stuart (1997), *Thinking with Demons: The Idea of Witchcraft in Early Modern Europe*, Oxford: Clarendon Press.

Clastres, Pierre, and Claude Lefort (1976), 'Appendice à Étienne de La Boétie', in *Le discours de la servitude volontaire*, Paris: Payot, 229–307.

Closson, Marianne (2000), *L'imaginaire démoniaque en France*, Geneva: Droz.

Clucas, Stephen (ed.) (2006), *John Dee: Interdisciplinary Studies in English Renaissance Thought*, Dordrecht: Springer.

Clucas, Stephen (2011), *Magic, Memory and Natural Philosophy in the Sixteenth and Seventeenth Centuries*, Aldershot: Ashgate.

Cohen, Adam Max (2009), *Technology and the Early Modern Self*, New York: Palgrave Macmillan.

Cohen, Adam Max (2012), 'Science and Technology', in Arthur F. Kinney (ed.), *The Oxford Handbook of Shakespeare*, Oxford: Oxford University Press, 702–19.

Colie, Rosalie L. (1955), 'Cornelis Drebbel and Salomon de Caus: Two Jacobean Models for Salomon's House', *Huntington Library Quarterly*, 18.3 (May), 245–60.

Collini, Stefan (1993), 'Introduction', in C. P. Snow, *The Two Cultures*, Cambridge: Cambridge University Press.

Corfield, Cosmo (1985), 'Why Did Prospero Abjure his Rough Magic?', *Shakespeare Quarterly*, 36.1 (Spring), 31–48.

Cortanze, Gérard de (1987), *Le Baroque*, Paris: MA Éditions.

Cottegnies, Line (ed. and trans.) (2008), *1 Henri VI*, in *Histoires*, vol. 1, ed. Jean-Michel Déprats and Gisèle Venet, Paris: Gallimard, 'Bibliothèque de la Pléiade'.

Coursen, H. R. (1999), 'Stoppard's *Rosencrantz and Guildenstern Are Dead*: The Film', in Lois Potter and Arthur F. Kinney (eds), *Shakespeare, Text and Theater: Essays in Honour of Jay L. Halio*, Newark: University of Delaware Press, 183–93.

Cox, Charles J. (2008) [1923], *English Church Fittings, Furniture and Accessories*, Huddersfield: Jeremy Mills Publishing.

Craik, T. W. (ed.) (1990), *The Merry Wives of Windsor*, The Oxford Shakespeare, Oxford: Oxford University Press.

Crane, Mary Thomas (2013), 'Optics', in Henry S. Turner (ed.), *Early Modern Theatricality*, Oxford: Oxford University Press, 250–69.

Crane, Mary Thomas (2014), *Losing Touch with Nature: Literature and the New Science in Sixteenth-Century England*, Baltimore: Johns Hopkins University Press.

Cressy, David, and Lori Anne Ferrell (1996), *Religion and Society in Early Modern England: A Sourcebook*, 2nd edn, New York: Routledge.

Crombie, A. C. (1990), *Science, Optics and Music in Medieval and Early Modern Thought*, London: Hambledon Press.

Crone, Hugh (2004), *Paracelsus, the Man who Defied Medicine: His Real Contribution to Medicine and Science*, Melbourne: Albarello Press.

Cusset, François (2003), *French Theory: Foucault, Derrida, Deleuze & Cie et les mutations de la vie intellectuelle aux États-Unis*, Paris: La Découverte.

Dandrey, Patrick (1997), 'Montaigne paradoxal', in *L'éloge paradoxal de Gorgias à Molière*, Paris: Presses Universitaires de France, 137–73.

Daston, Lorraine, and Katherine Park (1998), *The Cambridge History of Science*, Cambridge: Cambridge University Press, vol. 3.

Dawbarn, Frances, and Stephen Pumfrey (2004), 'Science and Patronage in England, 1570–1625', *History of Science*, 42, 137–88.

Dear, Peter (1995), *Discipline and Experience: The Mathematical Way in the Scientific Revolution*, Chicago: University of Chicago Press.

Dear, Peter (2001), *Revolutionizing the Sciences: European Knowledge and its Ambitions, 1500–1700*, Princeton: Princeton University Press.

Debord, Guy (1983), *Society of the Spectacle*, Detroit: Black and Red.

Debus, Allen G., and Michael T. Walton (eds) (1997), *Reading the Book of Nature: The Other Side of the Scientific Revolution*, Kirksville: Thomas Jefferson University Press.

Defaux, Gérard (1987), *Marot, Rabelais, Montaigne: l'écriture comme présence*, Paris: Champion.

Del Sapio Garbero, Maria (2010), 'A Spider in the Eye/I: The Hallucinatory Staging of the Self in Shakespeare's *The Winter's Tale*', in Ute Berns (ed.), *Solo Performances: Staging the Early Modern Self in England*, Amsterdam and New York: Rodopi, 133–55.

Del Sapio Garbero, Maria (2011), 'Troubled Metaphors: Shakespeare and the Renaissance Anatomy of the Eye', in Klaus Bergdolt and Manfred Pfister (eds), *Dialoge zwischen Wissenschaft, Kunst und Literatur in der Renaissance*, Wolfenbütteler Abhandlungen zur Renaissance-forschung, 27, Wiesbaden: Harrassowitz, 43–70.

DiGangi, Mario (ed.) (2008), *The Winter's Tale: Texts and Context*, Boston and Bedford: St Martin's.

Digges, Leonard (1555) [1553], *A prognostication of right good effect, fructfully augmented, contayninge playne, briefe, pleasant, chosen rules, to iudge the wether for euer, by the sunne, moone, sterres, cometes, raynbowe, thunder, cloudes, with other extraordinarie tokens, not omitting the aspectes of planetes, with a brefe iudgemente for euer, of plentie, lacke, sickenes, death, vvarres &c.*, Imprynted at London, Within the blacke Fryars, by Thomas Gemini, STC (2nd edn) 435.35.

Digges, Leonard (1576), *A prognostication euerlastinge of right good effecte fruitfully augmented by the auctour, contayning plaine, briefe, pleasau[n]t, chosen rules to iudge the weather by the sunne, moone, starres, comets, rainebow, thunder, cloudes, with other extraordinary tokens, not omitting the aspects of planets, vvith a briefe iudgement for euer, of plenty, lacke, sickenes, dearth, vvarres &c. opening also many naturall causes vvorthy to be knovven*, Imprinted at London by Thomas Marsh, STC (2nd edn) 435.47.

Diogenes Laertius (1964), *Diogenis Laertii Vitae philosophorum*, ed. H. S. Long, Oxford: Clarendon.

Dionysius Lambinus (1563), *Titi Lucretii Cari De rerum natura libri sex*, Paris and Lyon: Rovillium (Guillaume Rouillé and Philippe Rouillé).

Dohrn-van Rossum, Gerhard (1996), *History of the Hour: Clocks and Modern Temporal Orders*, Chicago: University of Chicago Press.

Dowland, John (1600), *Second Booke of Ayres*, London.

Dowland, John (1604), *Lachrimae, or Seaven Tears*, London.

Drakakis, John (2002), *Alternative Shakespeares*, London: Routledge.

Drouet, Pascale (2009), 'Introduction', in Pascale Drouet (ed.), *The Spectacular in and around Shakespeare*, Newcastle: Cambridge Scholars.

Dulac, Anne-Valérie (2009), 'Shakespeare et l'optique arabe', in Pierre Kapitaniak and Jean-Michel Déprats (eds), *Actes des congrès de la Société française Shakespeare*, 27; available at <http://shakespeare.revues.org/1498> (last accessed 27 January 2017).

Duplessis-Mornay, Philippe (1587), *A woorke concerning the trewnesse of the Christian religion*, trans. Philip Sidney and Arthur Golding, London, [By [John Charlewood and] George Robinson] for Thomas Cadman, STC (2nd edn) 18,149.

Dusinberre, Juliet (2003), 'Pancakes and a Date for *As You Like It*', *Shakespeare Quarterly*, 54, 371–405.

Dutton, Richard (2016), *Shakespeare, Court Dramatist,* Oxford: Oxford University Press.

Eagleton, Terry (2004), *After Theory*, London: Penguin Books.

Eamon, William (2004), 'Astrology and Society', in Brendan Dooley (ed.), *A Companion to Astrology in the Renaissance*, Leiden and Boston: Brill, 141–92.

Eccles, Audrey (1982), *Obstetrics and Gynaecology in Tudor and Stuart England*, London and Canberra: Croom Helm.

Edwards, David L. (2001), *John Donne: Man of Flesh and Spirit*, London: Continuum.

Edwards, Paul (2001), 'Science in *Hapgood* and *Arcadia*', in Katherine E. Kelly (ed.), *The Cambridge Companion to Tom Stoppard*, Cambridge: Cambridge University Press, 171–84.

Egan, Gabriel (2006), *Green Shakespeare: From Ecopolitics to Ecocriticism*, London: Routledge.

Egan, Gabriel (2015), *Shakespeare and Ecocritical Theory*, London: Bloomsbury.

Elam, Keir (1993), 'After Magritte, After Carroll, After Wittgenstein: What Tom Stoppard's Tortoise Taught Us', in Hersh Zeifman and Cynthia Zimmerman (eds), *Contemporary British Drama, 1970–90: Essays from Modern Drama*, London: Macmillan, 184–98.

Erasmus (1964), *The Adages of Erasmus*, ed. M. Mann Phillips, Cambridge: Cambridge University Press.

Erasmus, Desiderius (2006) [1524], *The Alchemist*, in W. B. Jensen (ed.), *Bulletin for the History of Chemistry*, 31.2, 61–5.

Fale, Thomas (1593), *Horologiographia: The art of dialling*, London, Printed by Thomas Orwin, dwelling in Pater noster-Row ouer against the signe of the Checker, STC (2nd edn) 10,678.

Fattori, Marta (2012), *Études sur Francis Bacon*, Paris: Presses Universitaires de France.

Feingold, Mordechai (1984), *The Mathematicians' Apprenticeship: Science, Universities and Society in England, 1560–1640*, Cambridge: Cambridge University Press.

Festugière, André-Jean (1941), 'Le symbole du Phénix et le mysticism hermétique', *Monuments Piot*, 38, 147–51.

Ficino, Marsilio (1576), *In Timæus Commentarium*, in *Ficini Opera omnia*, Basileae, ex officina Henricpetrina.

Field, Judith Veronica (1997), *The Invention of Infinity: Mathematics and Art in the Renaissance*, Oxford and New York: Oxford University Press.

Fineman, Joel (1986), *Shakespeare's Perjured Eye: The Invention of Poetic Subjectivity in the Sonnets*, Los Angeles: University of California Press.

Fleming, James Dougal (2011), *The Invention of Discovery, 1500–1700*, Farnham: Ashgate.

Fleming, John (2001), *Stoppard's Theatre: Finding Order Amid Chaos*, Austin: University of Texas Press.

Florio, John (trans.) (1999) [1603], *Montaigne's Essays*, Renascence Editions e-text provided by B. R. Schneider, © 1999 The University of Oregon, <http://www.luminarium.org/renascence-editions/montaigne> (last accessed 27 January 2017).

Floyd-Wilson, Mary (2013), *Occult Knowledge, Science, and Gender on the Shakespearean Stage*, Cambridge: Cambridge University Press.

Foakes, R. A. (ed.) (2002), *Henslowe's Diary*, 2nd edn, Cambridge: Cambridge University Press.

Force, Pierre (1997), 'Beyond the Metalanguage: Bathmology', in Jean-Michel Rabaté (ed.), *Writing the Image after Roland Barthes*, Philadelphia: University of Pennsylvania Press, 187–95.

Fowler, Alastair (1970), *Triumphal Forms: Structural Patterns in Elizabethan Poetry*, Cambridge: Cambridge University Press.

Fox, Alice (1979), 'Obstetrics and Gynecology in *Macbeth*', *Shakespeare Studies*, 12, 127–41.

Fox, Robert (ed.) (2000), *Thomas Harriot: An Elizabethan Man of Science*, Aldershot: Ashgate.

Freeman, John (1996), 'Holding up the Mirror to Mind's Nature: Reading *Rosencrantz* "Beyond Absurdity"', *The Modern Language Review*, 91.1, 20–39.

French, Peter (1972), *John Dee: The World of an Elizabethan Magus*, London: Routledge & Kegan Paul.

Frisius, Gemma (1584), *Charta cosmographica*, in *Cosmographia, sive descriptio universi orbis*, Antwerp: Jan Verwithagen for Arnold Coninx.

Frye, Roland Mushat (1984), *The Renaissance Hamlet: Issues and Responses in 1600*, Princeton: Princeton University Press.

Fulke, William (1560), *Antiprognostication*, London.

Fumaroli, Marc (1972), 'Microcosme comique et macrocosme solaire: Molière, Louis XIV, et *L'Impromptu de Versailles*', *Revue des sciences humaines*, 37.145 (January–March), 95–114.

Fumée, Marin (trans.) (1587), *Histoire générale des Indes Occidentales*, Paris.

Furniss, W. Todd (1954), 'The Annotation of Ben Jonson's *Masqve of Qveenes*', *The Review of English Studies*, New Series, 5.20, 344–60.

Galilei, Galileo (2001), *Dialogue Concerning the Two Chief World Systems*, trans. Stillman Drake, New York: Modern Library.

Garber, Marjorie (2002), 'Post-Shakespearean Writers and the Transmission of Shakespeare Roman Numerals', in Thomas Moisan and Douglas Bruster (eds), *In the Company of Shakespeare: Essays on English Renaissance Literature in Honor of G. Blakemore Evans*, Cranbury: Associated University Presses, 223–50.

Garber, Marjorie (2005), *Shakespeare After All*, New York: Anchor Books.

García-Ballester, Luis (2002), *Galen and Galenism: Theory and Medical Practice from Antiquity to the European Renaissance*, ed. Jon Arrizabalaga,

Montserrat Cabré, Lluís Cifuentes and Fernando Salmón, Variorum Collected Studies Series, Aldershot: Ashgate.

Gardette, Raymond (1994), '"Some vanity of mine art": l'espace magique dans *La tempête*', in *Shakespeare, La tempête. Études critiques; Actes du Colloque de Besançon, 2–3–4 décembre 1993*, Besançon: University of Franche-Comté, 21–37.

Gaskill, Malcolm (2006), *Witchfinders: A Seventeenth-Century English Tragedy*, London: John Murray Publishers.

Gerschow, Frederic (1892), *Travel Diary, September–October, 1602*, in *Transactions of the Royal History Society*, 6, 1–68.

Gibson, Kirsten (2009), 'Music, Melancholy and Masculinity in Early Modern England', in Ian Biddle and Kirsten Gibson (eds), *Masculinity and Western Musical Practice*, Farnham: Ashgate, 41–66.

Gibson, Marion (2014), *Shakespeare's Demonology: A Dictionary*, London: Bloomsbury.

Giglioni, Guido, James A. T. Lancaster, Sorana Corneanu and Dana Jalobeanu (eds) (2016), *Francis Bacon on Motion and Power*, Dordrecht: Springer.

Gillespie, Stuart (2007), 'Lucretius in the English Renaissance', in Stuart Gillespie and Philip Hardie (eds), *The Cambridge Companion to Lucretius*, Cambridge: Cambridge University Press, 242–53.

Gilman, Sander L., Helen King, Roy Porter, G. S. Rousseau and Elaine Showalter (1993), *Hysteria Beyond Freud*, Berkeley: University of California Press.

Gingerich, Owen (1981), 'Astronomical Scrapbook: Great Conjunctions, Tycho and Shakespeare', *Sky and Telescope*, 61, 39–95.

Glennie, Paul, and Nigel Thrift (2009), *Shaping the Day*, Oxford: Oxford University Press.

Golding, Arthur (trans.) (2002) [1567], *Ovid's Metamorphoses, Translated by Arthur Golding*, ed. Madeleine Forey, Baltimore: Johns Hopkins University Press.

Gordon, C. A. (1962), *A Bibliography of Lucretius*, London: Hart-Davis.

Gosson, Stephen (1895) [1579], *The Schoole of Abuse*, E. Arber's English Reprints, Westminster.

Gouk, Penelope (2000), 'Music, Melancholy and Medical Spirits in Early Modern Thought', in Peregrine Horden (ed.), *Music as Medicine: The History of Music Therapy since Antiquity*, Aldershot: Ashgate, 173–94.

Grant, Roger Matthew (2014), *Beating Time and Measuring Music in the Early Modern Era*, Oxford: Oxford University Press.

Graziani, René (1984), 'The Numbering of Shakespeare's Sonnets: 12, 60, and 126', *Shakespeare Quarterly*, 35.1 (Spring), 79–82.

Greenblatt, Stephen (1993), 'Shakespeare Bewitched', in Jeffrey N. Cox and Larry J. Reynolds (eds), *New Historical Literary Study: Essays on Reproducing Texts, Representing History*, Princeton: Princeton University Press, 108–35.

Greenblatt, Stephen (2001), *Hamlet in Purgatory*, Princeton: Princeton University Press.

Greenblatt, Stephen (2005) [1980], *Renaissance Self-Fashioning: From More to Shakespeare*, Chicago and London: University of Chicago Press.

Greenblatt, Stephen (2011), *The Swerve: How the World Became Modern*, New York: W. W. Norton & Company.

Greenblatt, Stephen, and Peter G. Platt (2014), *Shakespeare's Montaigne: The Florio Translation of the Essays: A Selection*, New York: New York Review Books.

Gurr, Andrew (2004), *Playgoing in Shakespeare's London*, Cambridge: Cambridge University Press.

Hackett, Helen (2007), 'The Rhetoric of (In)fertility: Shifting Responses to Elizabeth I's Childlessness', in Jennifer Richards and Alison Thorne (eds), *Rhetoric, Women and Politics in Early Modern England*, New York: Routledge, 149–71.

Hakluyt, Richard (1589), *The Principall Navigations, Voiages and Discoveries of the English Nation, made by Sea or over Land, to the most remote and farthest distant Quarters of the earth*, London: George Bishop and Ralph Newberie.

Halliwell-Phillipps, James Orchard (ed.) (1855), *The Works of William Shakespeare*, London: J. E. Adlard.

Hallyn, Fernand (2008), *Gemma Frisius, arpenteur de la terre et du ciel*, Paris: Champion.

Harding, D. W. (1969), 'Women's Fantasy of Manhood: A Shakespearian Theme', *Shakespeare Quarterly*, 20, 245–53.

Harkness, Deborah E. (1999), *John Dee's Conversations with Angels: Cabal, Alchemy, and the End of Nature*, Cambridge: Cambridge University Press.

Harkness, Deborah E. (2002), '"Strange" Ideas and "English" Knowledge', in Pamela H. Smith and Paula Findlen (eds), *Merchants and Marvels: Commerce, Science, and Art in Early Modern Europe*, London: Routledge, 137–62.

Harkness, Deborah E. (2007), *The Jewel House: Elizabethan London and the Scientific Revolution*, New Haven and London: Yale University Press.

Harlow, C. G. (1965), 'The Authorship of *1 Henry VI* (Continued)', *Studies in English Literature, 1500–1900*, 5.2, 269–81.

Harvey, Elizabeth D. (1993), *Ventriloquized Voices: Feminist Theory and English Renaissance Texts*, New York: Routledge.

Hattaway, Michael (ed.) (1990), *1 Henry VI*, The New Cambridge Shakespeare, Cambridge: Cambridge University Press.

Hawkes, David (2014), 'Proteus Agonistes: Shakespeare, Bacon, and the "Torture" of Nature', in Laurie Johnson, John Sutton and Evelyn Tribble (eds), *Embodied Cognition and Shakespeare's Theatre: The Early Modern Body-Mind*, New York: Routledge, 13–26.

Hayman, Ronald (1979), *Tom Stoppard*, 3rd edn, London: Heinemann.

Healy, Margaret (2011), *Shakespeare's Alchemy and the Creative Imagination: The Sonnets and A Lover's Complaint*, Cambridge: Cambridge University Press.

Heisenberg, Werner (1983), 'The Physical Content of Quantum Kinematics', trans. J. A. Wheeler and W. H. Zurek, in John A. Wheeler and Wojciech H. Zurek (eds), *Quantum Theory and Measurement*, Princeton: Princeton University Press, 62–84.

Herbert, George (1841) [1640], *Jacula Prudentum or, Outlandish Proverbs, Sentences etc*, in *The Remains of that sweet singer of The Temple, George Herbert*, London: Pickering.

Hibbard, G. R. (ed.) (1990), *Love's Labour's Lost*, The Oxford Shakespeare, Oxford: Clarendon Press.

Hibbard, G. R. (ed.) (1994) [1987], *Hamlet*, Oxford: Oxford World's Classics.

Hickey, Helen (2015), 'Medical Diagnosis and the Colour Yellow in Early Modern England', *E-rea*, ed. Sophie Chiari, 12.2, <http://erea.revues.org/4413> (last accessed 27 January 2017).

Hillman, David (2007), *Shakespeare's Entrails: Belief, Skepticism and the Interior of the Human Body*, New York: Palgrave Macmillan.

Hillman, David, and Carla Mazzio (eds) (1997), *The Body in Parts: Fantasies of Corporeality in Early Modern Europe*, New York: Routledge.

Hippocrates (1975), *Hippocrates: Diseases of Women 1*, trans. Ann Ellis Hanson, *Signs*, 1.2, 567–84.

Hoeniger, David (1992), *Medicine and Shakespeare in the English Renaissance*, Newark: University of Delaware Press.

Holland, Peter (ed.) (1994), *A Midsummer Night's Dream*, Oxford: Oxford World's Classics.

Hollander, John (1961), *The Untuning of the Sky: Ideas of Music in English Poetry, 1570–1700*, Princeton: Princeton University Press.

Höltgen, Karl Josef (2004), 'Quarles, Francis (1592–1644)', in *Oxford Dictionary of National Biography*, Oxford: Oxford University Press, <http://www.oxforddnb.com/view/article/22945> (last accessed 27 January 2017).

Horden, Peregrine (ed.) (2000), *Music as Medicine: The History of Music Therapy since Antiquity*, Aldershot: Ashgate.

Houston Wood, David (2002), '"He Something Seems Unsettled": Melancholy, Jealousy, and Subjective Temporality in *The Winter's Tale*', *Renaissance Drama*, 31, 185–213.

Howard, Henry (1583), *A defensative against the poison of supposed prophesies*, London: John Charlewood.

Howard Traister, Barbara (1984), *Heavenly Necromancers: The Magician in English Renaissance Drama*, Columbia: University of Missouri Press.

Hunter, Robert G. (1978), 'It Strikes Us Today: Falstaff and the Protestant Ethic', in David Bevington and Jay L. Halio (eds), *Shakespeare: Pattern of Excelling Nature*, Cranbury: Associated University Presses, 125–32.

Irving, I. Edgar (1970), *Shakespeare, Medicine and Psychiatry: An Historical Study in Criticism and Interpretation*, New York: Philosophical Library.

Iselin, Pierre (1995a), 'Music and Difference: Elizabethan Stage Music and its Reception', in Jean-Marie Maguin and Michèle Willems (eds), *French Essays on Shakespeare and his Contemporaries: 'What would France with us?'*, Newark: University of Delaware Press, 96–113.

Iselin, Pierre (1995b), '"My Music for Nothing": Musical Negotiations in *The Tempest*', *Shakespeare Survey*, 48, 135–45.

Iselin, Pierre (2014a), 'The Apology of Music in England (1579–1605): The Paradox of a Polemical *Topos*', in Sophie Chiari and Hélène Palma (eds), *Transmission and Transgression: Cultural Challenges in Early Modern England*, Aix-en-Provence: Presses Universitaires de Provence, 63–77.

Iselin, Pierre (2014b), 'Commentary on "O Mistress Mine", no. 7: An Introduction to the Audio CD "Shakespeare in Music"', in Sophie Chiari and Hélène Palma (eds), *Transmission and Transgression: Cultural Challenges in Early Modern England*, Aix-en-Provence: Presses Universitaires de Provence, 167–8.

James VI (1924) [1597], *Dæmonologie and Newes from Scotland*, ed. G. B. Harrison, London: The Bodley Head.

J.C. (1620), *The Two Merry Milkmaids*, London: Bernard Alsop.

Jernigan, Daniel Keith (2012), *Tom Stoppard: Bucking the Postmodern*, Jefferson: McFarland.

Johnson, Francis R. (1968) [1937], *Astronomical Thought in Renaissance England*, New York: Octagon Books.

Johnson, Francis R., and S. V. Larkey (1934), 'Thomas Digges, the Copernican System and the Idea of the Infinity of the Universe', *The Huntington Library Bulletin*, 5, 69–117.

Johnston, Stephen (2004a), 'Digges, Leonard (*c.* 1515–*c.* 1559)', in *Oxford Dictionary of National Biography*, Oxford: Oxford University Press, <http://www.oxforddnb.com/view/article/7637> (last accessed 27 January 2017).

Johnston, Stephen (2004b), 'Digges, Thomas (*c.* 1546–1595)', in *Oxford Dictionary of National Biography*, Oxford: Oxford University Press, <http://www.oxforddnb.com/view/article/7639> (last accessed 27 January 2017).

Jones-Davies, Margaret (1986), 'Paroles intertextuelles: lecture intertextuelle de Parolles', in Jean Fuzier and François Laroque (eds), *Actes du Colloque: All's Well that Ends Well*, Montpellier: Collection Astrea, 65–80.

Jones-Davies, Margaret (1993), 'L'échiquier et Médée: deux points de controverse dans *La Tempête*', *Études anglaises*, 4, 447–51.

Jones-Davies, Margaret (2003), '*Cymbeline* and the Sleep of Faith', in R. Dutton, A. Findlay and R. Wilson (eds), *Theatre and Religion: Lancastrian Shakespeare*, Manchester: Manchester University Press, 197–217.

Jones-Davies, Margaret (2005a), 'Defacing the Icon of the King: *Richard II* and the Issues of Iconoclasm', in Guillaume Winter (ed.), *Autour de Richard II de William Shakespeare*, Arras: Artois Presses Université, 101–12.

Jones-Davies, Margaret (2005b), '*Richard II* as *ludus puerorum* and the Perfecting of the King', in Isabelle Schwartz-Gastine (ed.), *Richard II de William Shakespeare, une oeuvre en contexte*, Maison de la Recherche en Sciences Humaines de Caen, Special Issue, 35–55.

Jones-Davies, Margaret (2010), '"Suspension of disbelief" in *The Winter's Tale*', *Études anglaises*, 63.3 (July–September), 259–73.

Jones-Davies, Margaret (2014), 'The Process of Secularization in the Renaissance: Shakespeare and Modern Evil', in Sophie Chiari and Hélène Palma (eds), *Transmission and Transgression: Cultural Challenges in Early Modern England*, Aix-en-Provence: Presses Universitaires de Provence, 39–49.

Jones-Davies, Margaret (2015), 'The Chess Game and Prospero's Epilogue in *The Tempest*', *Notes and Queries*, 260.1 (March), 118–20.

Jonson, Ben (1963) [1941], *Ben Jonson*, ed. C. H. Herford and Percy and Evelyn Simpson, 10 vols, Oxford: Clarendon Press.

Jonson, Ben (1967), *The Alchemist*, ed. F. H. Mares, London: Methuen.

Jorden, Edward (1991) [1603], *A Briefe Discourse of a Disease Called the Suffocation of the Mother*, in Michael Macdonald, *Witchcraft and Hysteria in Elizabethan London: Edward Jorden and the Mary Glover Case*, London: Routledge.

Jung, C. G. (1970) [1943], *Psychologie et alchimie*, ed. and trans. Henry Pernet and Roland Cahen, Paris: Buchet/Chastel.

Kantorowicz, Ernst (1957), *The King's Two Bodies: A Study in Medieval Political Theology*, Princeton: Princeton University Press.

Kapitaniak, Pierre (2007), 'Entre ruse diabolique et illusion dramatique: cheminements du discours démonologique dans l'Angleterre jacobéenne', in Françoise Lavocat, Pierre Kapitaniak and Marianne Closson (eds), *Fictions du diable*, Geneva: Droz.

Kapitaniak, Pierre (2008), *Spectres, ombres et fantômes: discours et représentations dramatiques en Angleterre, 1576–1642*, Paris: Honoré Champion.

Kargon, R. H. (1966), *Atomism in England from Hariot to Newton*, Oxford: Clarendon Press.

Kassell, Lauren (2005), *Medicine and Magic in Elizabethan London. Simon Forman: Astrologer, Alchemist, and Physician*, Oxford: Oxford University Press.

Kermode, Frank (1966), 'Introduction', in William Shakespeare, *The Tempest*, ed. Frank Kermode, Arden Shakespeare Paperback, London: Methuen.

Kernan, Alvin B. (1979), *The Playwright as Magician: Shakespeare's Image of the Poet in the English Public Theater*, New Haven and London: Yale University Press.

Kerwin, William (2005), *Beyond the Body: The Boundaries of Medicine and English Renaissance Drama*, Amherst: University of Massachusetts Press.

Kiefer, Frederick (2003), *Shakespeare's Visual Theatre: Staging the Personified Characters*, Cambridge: Cambridge University Press.

Kinney, Arthur F. (2004), *Shakespeare's Webs: Networks of Meaning in Renaissance Drama*, New York: Routledge.

Koyré, Alexandre (1957), *From the Closed World to the Infinite Universe*, Baltimore: Johns Hopkins University Press.

La Belle, Jenijoy (1980), '"A Strange Infirmity": Lady Macbeth's Amenorrhea', *Shakespeare Quarterly*, 31, 381–6.

La Boétie, Étienne (2012), *Discourse on Voluntary Servitude*, trans. James B. Atkinson and David Sices, introduction and notes by James B. Atkinson, Indianapolis: Hackett Publishing Company.

Lafond, Jean (1984), 'Le *Discours de la servitude volontaire* de La Boétie et la rhétorique de la déclamation', in Pierre-Georges Castex (ed.), *Mélanges sur la littérature de la Renaissance à la mémoire de V.-L. Saulnier*, Geneva: Droz, 735–45.

Laoutaris, Chris (2008), *Shakespearean Maternities: Crises of Conception in Early Modern England*, Edinburgh: Edinburgh University Press.

Laroque, François (1991), *Shakespeare's Festive World*, Cambridge: Cambridge University Press.

Laroque, François (1995), 'Tradition and Subversion in *Romeo and Juliet*', in Jay L. Halio (ed.), *Shakespeare's Romeo and Juliet: Texts, Contexts, and Interpretation*, Newark: University of Delaware Press; London: Associated University Presses, 18–36.

Laroque, Fançois (ed.) (1997), *Le Docteur Faust/Doctor Faustus*, trans. François Laroque and Jean-Pierre Villquin, Paris: GF Flammarion.

Laroque, François (ed. and trans.) (2011), *La Tempête*, Le Livre de Poche, Paris: LGF.

Laurens, Pierre (1985), 'Monsieur de Montaigne disait que . . . (bathmologies)', in Frank Lestringant (ed.), *Rhétorique de Montaigne*, Paris: Champion, 73–84.

Lavater, Ludwig (1572), *Of ghostes and spirites walking by nyght*, London: Henry Bynnemann for Richard Watkins.

Le Goff, Jacques (1999), 'Temps', in Jacques Le Goff and Jean-Claude Schmitt (eds), *Dictionnaire raisonné de l'Occident médiéval*, Paris: Fayard, 1, 113–22.

Le Testu, Guillaume (2012) [1555], *Cosmographie universelle. Selon les navigateurs tant anciens que modernes par Guillaume Le Testu pillotte en la mer du Ponent, de la ville françoyse de Grace*, ed. Frank Lestringant, Paris: Arthaud, Direction de la Mémoire, du Patrimoine et des Archives, Carnets des Tropiques.

Lecercle, Ann (1991), 'The Letter That Killeth: The Desacralized and the Diabolic Body in Shakespeare', in M. T. Jones-Davies (ed.), *Shakespeare et le corps à la Renaissance*, Paris: Les Belles Lettres, 137–52.

Leon Alfar, Cristina (2003), *Fantasies of Female Evil: The Dynamics of Gender and Power in Shakespearean Tragedy*, Newark: University of Delaware Press.

Lepage, John L. (2012), *The Revival of Antique Philosophy in the Renaissance*, New York: Palgrave Macmillan.

Leppert, Richard (1993), 'Social Order and the Domestic Consumption of Music (The Politics of Sound in the Policing of Gender Construction)', in *The Sight of Sound: Music Representation and the History of the Body*, Berkeley, Los Angeles and London: University of California Press, 63–90 (chapter reprinted in Ann Bermingham and John Brewer (eds) (1995), *The Consumption of Culture, 1600–1800: Image, Object, Text*, New York: Routledge).

Leslie, John (1572), *A Treatise of Treasons against Q. Elizabeth*, London: J. Fowler.

Lestringant, Frank (1994), *Le Cannibale, grandeur et decadence*, Paris: Perrin.

Lestringant, Frank (2004), 'Gonzalo's Books: La République des Cannibales, de Montaigne à Shakespeare', in Pierre Kapitaniak and Jean-Marie Maguin (eds), *Shakespeare et Montaigne. Vers un nouvel humanisme. Actes du Congrès organisé par la Société française Shakespeare en collaboration avec la Société internationale des Amis de Montaigne (13–15 mars 2003)*, Montpellier: AVL Diffusion, 175–93.

Levak, P. Brian (ed.) (2004), *The Witchcraft Sourcebook*, New York: Routledge.

Levenson, Jill L. (ed.) (2000), *Romeo and Juliet*, The Oxford Shakespeare, Oxford: Oxford University Press.

Lévi-Strauss, Claude (1991), *Histoire de lynx*, Paris: Plon.

Lévi-Strauss, Claude (1996), *The Story of Lynx*, trans. Catherine Tihanyi, Chicago: University of Chicago Press.

Levin, Harry (1970), *The Myth of the Golden Age in the Renaissance*, London: Faber and Faber.

Levin, Joanna (2002), 'Lady Macbeth and the Daemonology of Hysteria', *English Literary History*, 69.1, 21–55.

Lewens, Tim (2015), *The Meaning of Science: A Pelican Introduction*, London: Penguin.

Libera, Alain de, Claude Gauvard and Michel Zink (eds) (2002), *Dictionnaire du Moyen Âge*, Paris: Presses Universitaires de France.

Lim, Walter S. H. (2001), 'Knowledge and Belief in *The Winter's Tale*', in *Studies in English Literature*, 41.2, 317–34.

Lindberg, David (ed.) (1970), *John Pecham and the Science of Optics: Perspectiva Communis*, Madison: University of Wisconsin Press.

Lindberg, David (2002), 'Alhacen', in Randall Curren (ed.), *A Companion to Philosophy in the Middle Ages*, Malden and Oxford: Blackwell, 127–8.

Lindley, David (2006), *Shakespeare and Music*, Arden Critical Companions, The Arden Shakespeare, London: Thomson Learning.

Llull, Ramon (1987), *Le Livre de l'Ami et de l'Aimé*, trans. Max Jacob, Paris: Fata Morgana.

Lobanov-Rostovsky, Sergei (2010), 'Taming the Basilisk', in David Hillman and Carla Mazzio (eds), *The Body in Parts: Fantasies of Corporeality in Early Modern Europe*, New York and Abingdon: Routledge, 195–219.

Lopez de Gomara, Francisco (1552), *Historia General de las Indias*.

Lucretius (1866), *Titi Lucreti Cari De rerum natura libri sex with notes and a translation*, ed. H. A. J. Munro, 2nd edn, 2 vols, Cambridge: Deighton Bell and Co.

Lucretius (1947), *Titi Lucreti Cari De rerum natura libri sex*, ed. Cyril Bailey, 3 vols, Oxford: Clarendon Press.

Lucretius (1955), *On the Nature of the Universe*, trans. R. Latham, Harmondsworth: Penguin Classics.

Lucretius (1998), *De la nature/De rerum natura*, trans. José Kany-Turpin, Paris: Flammarion.

Lund, Mary Ann (2010), *Melancholy, Medicine and Religion in Early Modern England: Reading the Anatomy of Melancholy*, Cambridge: Cambridge University Press.

Lyly, John (2000), *Galatea and Midas*, ed. George K. Hunter and David Bevington, Manchester: Manchester University Press.

Lyons, Bridget Gellert (1971), *Voices of Melancholy: Studies in Literary Treatments of Melancholy in Renaissance England*, London: Routledge & Kegan Paul.

McCanles, Michael (2005), 'The New Science and the *Via Negativa*: A Mystical Source for Baconian Empiricism', in Julie R. Solomon and Catherine Gimelli Martin (eds), *Francis Bacon and the Refiguring of Early Modern Thought: Essays to Commemorate The Advancement of Learning (1605–2005)*, Aldershot: Ashgate, 45–68.

McConnell, Anita (2004), 'Barlow, William (1544–1625)', in *Oxford Dictionary of National Biography*, Oxford: Oxford University Press, <http://www.oxforddnb.com/view/article/1444> (last accessed 27 January 2017).

Macdonald, Michael (1991), *Witchcraft and Hysteria in Elizabethan London: Edward Jorden and the Mary Glover Case*, London: Routledge.

Machielsen, Jan (2015), *Martin Delrio: Demonology and Scholarship in the Counter-Reformation*, Oxford: Oxford University Press.

MacKinnon, Dolly (2015), '"Ringing of Bells by Four White Spirits": Two Seventeenth-Century English Earwitness Accounts of the Supernatural in Print Culture', in Jennifer Spinks and Dagmar Eichberger (eds), *Religion, the Supernatural and Visual Culture in Early Modern Europe*, Leiden: Brill, 83–104.

Macrobius (1990), *Commentary on the Dream of Scipio*, ed. William Harris Stahl, New York: Columbia University Press.

Maisano, Scott (2004), 'Shakespeare's Last Act: The Starry Messenger and the Galilean Book in *Cymbeline*', *Configurations*, 12.3, 401–34.

Marchitello, Howard (2006), 'Science Studies and English Renaissance Literature', *Literature Compass*, 3.3, 341–65.

Marchitello, Howard (2011), *The Machine in the Text: Science and Literature in the Age of Shakespeare and Galileo*, Oxford: Oxford University Press.

Marouby, Christian (1990), 'Rhétorique de la négativité', in *Utopie et primitivisme. Essai sur l'imaginaire anthropologique à l'âge classique*, Paris: Éditions du Seuil.

Mason Vaughan, Virginia, and Alden T. Vaughan (eds) (2001), *The Tempest*, Arden Third Series, London: Thomson Learning.

Maurer, Armand (1962), *Medieval Philosophy*, New York: Random House.

Mazer, Arthur (2010), *The Ellipse: A Historical and Mathematical Journey*, Hoboken: Wiley.

Mazzio, Carla (ed.) (2009), 'Shakespeare and Science', *Special Double Issue of the Johns Hopkins Journal*, South Central Review (Winter and Spring).

Mebane, John S. (1989), *Renaissance Magic and the Return of the Golden Age: The Occult Tradition and Marlowe, Jonson and Shakespeare*, Lincoln: University of Nebraska Press.

Ménard, Louis (ed.) (1977), *Hermès Trismégiste*, Paris: Guy Trédaniel, Éditions de la Maisnie.

Middleton, Thomas (2012), *The Witch/La sorcière*, ed. Pierre Kapitaniak, Paris: Classiques Garnier.

Montaigne (1999) [1965], *Essais*, ed. Pierre Villey, Paris: PUF, Quadrige.

Montaigne (2001), *Les Essais*, ed. Jean Céard et al., La Pochothèque, Paris: LGF.

Morley, Thomas (1963) [1597], *A Plain and Easy Introduction to Practical Music*, ed. R. Alec Harman, London and New York: W. W. Norton, revised edn.

Mortimer, Ian (2012), *The Time Traveller's Guide to Elizabethan England*, London: The Bodley Head.

Mowat, Barbara (1981), 'Prospero, Agrippa and Hocus Pocus', *English Literary Renaissance*, 11, 281–303.

Mowat, Barbara (2001), 'Prospero's Book', *Shakespeare Quarterly*, 52.1, 1–33.

Muir, Kenneth (1977), *The Sources of Shakespeare's Plays*, London: Methuen.

Muñoz Simonds, Peggy (1986), 'Eros and Anteros, in Shakespeare's Sonnets, 153 and 154: An Iconographical Study', in *Spenser Studies*, 7, 261–325.

Nardo, Anna K. (2008), 'Stoppard's Space Men: Rosencrantz and Guildenstern on Film', *Literature-Film Quarterly*, 36.2, 113–21.

Nashe, Thomas (1592), *The Apologie of Pierce Pennilesse*, London: Abell Jeffes.

Nashe, Thomas (1594), *The Terrors of the Night*, London: John Danter for William Jones.

Nashe, Thomas (1596), *Have with you to Saffron-Walden*, London: John Danter.

Neill, Michael (1997), *Issues of Death: Mortality and Identity in English Renaissance Tragedy*, Oxford: Clarendon Press.

Nicholas, Thomas (trans.) (1578), *The Pleasant History of the Conquest of the West Indias*, London.

Nicholson, Brinsley (ed.) (1886), *The Discoverie of Witchcraft, by Reginald Scot [. . .] Being a Reprint of the 1st Edition Published in 1584*, London: E. Stock.

Noble, Louise (2004), 'The *Fille Vièrge* as Pharmakon: The Therapeutic Value of Desdemona's Corpse', in Stephanie Moss and Kaara L. Peterson (eds), *Disease, Diagnosis and Cure on the Early Modern Stage*, Aldershot: Ashgate, 135–50.

Northbrooke, John (1577), *A Treatise wherein Dicing, Dauncing, Vaine Playes or Enterludes, with Idle Pastimes*, London.

Nuttall, A. D. (2007), *Shakespeare the Thinker*, New Haven: Yale University Press.

Ockham, William of (1974), *Summa Logicae Part I*, in *Ockham's Theory of Terms*, ed. Michael J. Loux, Notre Dame: University of Notre Dame Press.

Orgel, Stephen (1989), 'Nobody's Perfect: Or Why Did the English Stage Take Boys for Women?', *South Atlantic Quarterly*, 88.1, 7–29.

Orgel, Stephen (2011), *Spectacular Performances: Essays on Theatre, Imagery, Books and Selves in Early Modern England*, Manchester: Manchester University Press.

Orgel, Stephen, and Sean Keilen (eds) (1999), *Shakespeare: The Critical Complex, Vol X: Postmodern Shakespeare*, New York and London: Garland.

Ovid, *Metamorphoses* (1955), trans. M. Innes, Harmondsworth: Penguin Classics.

Pafford, J. H. P. (ed.) (1963), *The Winter's Tale*, The Arden Shakespeare, Walton-on-Thames: Thomas Nelson and Sons.

Paster, Gail Kern (1993), *The Body Embarrassed: Drama and the Disciplines of Shame in Early Modern England*, Ithaca: Cornell University Press.

Paster, Gail Kern (2004), *Humoring the Body: Emotions and the Shakespearean Stage*, Chicago: University of Chicago Press.

Patterson, W. B. (2000), *King James VI and I and the Reunion of Christendom*, Cambridge: Cambridge University Press.

Paul, Henry N. (1950), *The Royal Play of Macbeth: When, Why and How It Was Written by Shakespeare*, New York: Macmillan.

Peacham, Henry (1612), *The Gentlemans Exercise [. . .] in Lyming, Painting, etc.*, London.

Percy, William (2006) [c. 1601], *Mahomet and His Heaven*, ed. M. Dimmock, Aldershot: Ashgate.

Perez-Ramos, Antonio (1988), *Francis Bacon's Idea of Science and the Maker's Knowledge Tradition*, Oxford: Clarendon Press.

Pesic, Peter (1999), 'Wrestling with Proteus: Francis Bacon and the Torture of Nature', *Isis*, 90, 81–94.

Peterson, Kaara L. (2010), *Popular Medicine, Hysterical Disease and Social Controversy in Shakespeare's England*, Farnham: Ashgate.

Pettigrew, Todd Howard James (2007), *Shakespeare and the Practice of Physic: Medical Narratives on the Early Modern English Stage*, Delaware: University of Delaware Press.

Pina Martins, José V. de (1988), 'Um Livro acerca da Loucoura (*La Pazzia*, 1540), e a sua discutivel originalidade', in *Homenaje a Eugenio Asensio*, Madrid: Editorial Gredos, 361–78.

Pina Martins, José V. de (1992), 'Modèles portugais et italiens de Montaigne', in *Montaigne et l'Europe. Actes du colloque international de Bordeaux (1992)*, Mont-de-Marsan: Éditions Interuniversitaires, 139–52.

Pitcher, John (ed.) (2005), *Cymbeline*, London: Penguin Books.

Plato (1946), *The Republic*, trans. Paul Shorey, London: Loeb Classical Library.

Platt, Peter G. (1997), *Reason Diminished: Shakespeare and the Marvellous*, Lincoln and London: University of Nebraska Press.

Polidore Vergil (1546), *De inventoribus rerum*, trans. Thomas Langley, London.

Pollnitz, Aysha (2015), *Princely Education in Early Modern Britain*, Cambridge: Cambridge University Press.

Pollock, Jonathan (2009), '*King Lear* in the Light of Lucretius: "nullam rem e nihilo"', in François Laroque, Pierre Iselin and Sophie [Chiari] Alatorre (eds), *'And that's true too': New Essays on King Lear*, Newcastle: Cambridge Scholars Publishing, 165–77.

Pollock, Jonathan (2013), '"Past fearing death": Epicurean Ethics in *Measure for Measure*', *Actualité de Mesure pour Mesure, Sillages critiques*, 15, <http://sillagescritiques.revues.org/2544> (last accessed 27 January 2017).

Pollock, Jonathan (2015), 'Shakespeare and the Atomist Heritage', in Sophie Chiari (ed.), *The Circulation of Knowledge in Early Modern Literature*, Farnham: Ashgate, 47–58.

Popelard, Mickaël (2009), 'Spectacular Science: A Comparison of Shakespeare's *The Tempest*, Marlowe's *Doctor Faustus* and Bacon's *New Atlantis*', in Pascale Drouet (ed.), *The Spectacular in and around Shakespeare*, Newcastle: Cambridge Scholars Publishing, 17–40.

Popelard, Mickaël (2010), *Francis Bacon. L'humaniste, le magicien, l'ingénieur*, Paris: Presses Universitaires de France.

Popper, Karl (1992), *The Logic of Scientific Discovery*, London: Routledge.

Popper, Karl (2000) [1963], *Conjectures and Refutations: The Growth of Scientific Knowledge*, London and New York: Routledge.

Preston, Claire (2015), 'Review of *Losing Touch with Nature: Literature and the New Science in Sixteenth-Century England*, ed. Mary Thomas Crane. Baltimore: Johns Hopkins University Press, 2014', *Renaissance Quarterly*, 68.4 (Winter), 1,525–6.

Prynne, William (1633), *Histrio-Mastix*, London.

Pumfrey, Stephen, Paolo Rossi and Maurice Slawinski (eds) (1991), *Science, Culture and Popular Belief in Renaissance Europe*, Manchester: Manchester University Press.

Purkiss, Diane (1996), *The Witch in History: Early Modern and Twentieth-Century Representations*, London: Routledge.

Purkiss, Diane (2008), 'Fractious: Teenage Girls' Tales in and out of Shakespeare', in Mary Ellen Lamb and Karen Bamford (eds), *Oral Tradition and Genders in Early Modern Literary Texts*, Aldershot: Ashgate, 57–71.

Quignard, Pascal (1990), *Albucius*, Paris: P.O.L. and Livre de Poche.

Rabelais, François (1994), *Œuvres Complètes*, ed. Mireille Huchon with François Moreau, Paris: Gallimard, Bibliothèque de la Pléiade.

Racault, Jean-Michel (2010), 'Insularité, théâtralité et pouvoir dans *La Tempête* de Shakespeare', in *Robinson et compagnie. Aspects de l'insularité politique de Thomas More à Michel Tournier*, Paris: Éditions Petra (Des îles), 63–92.

Raleigh, Walter (1596), *The discoverie of the large, rich, and bewtiful empyre of Guiana*, London: R. Robinson.

Raleigh, Walter (1599), *Brevis et admiranda descriptio regni Guianae, auri abundantissimi, in America, seu Novo Orbe, sub linea aequinoctilia siti: Quod nuper admodum, annis nimirum 1594. 1595 et 1596. Per generosum dominum, Dn Gualtherum Ralegh equitem Anglum detectum est: paulo post jusssu ejus duobus libellis comprehensa*, Nuremberg: C. Lochner for Levinus Hulsius.

Ramsey, Jarold (1973), 'The Perversion of Manliness in *Macbeth*', *Studies in English Literature 1500–1900*, 13.2, 285–300.

Randall, John H. (1962), *Aristotle*, New York: Columbia University Press.

Reed, Robert Rentoul Jr (1952), *Bedlam on the Jacobean Stage*, Cambridge, MA: Harvard University Press.

Reilly, Kara (2011), *Automata and Mimesis on the Stage of Theatre History*, London: Palgrave Macmillan.

Rhodes, Neil (2014), 'Time', in Abigail Shinn, Matthew Dimmock and Andrew Hadfield (eds), *The Ashgate Research Companion to Popular Culture in Early Modern England*, Farnham: Ashgate, 283–94.

Richer, Jean (1990), *Lecture astrologique des pièces romaines de Shakespeare: Titus Andronicus, Jules César, Antoine et Cléopâtre, Coriolan*, Paris: Guy Trédaniel, Éditions de la Maisnie.

Rosencrantz and Guildenstern Are Dead (1990), dir. Tom Stoppard, Brandenburg International, DVD.

Rosenfield, Kristie Gulick (2002), 'Nursing Nothing: Witchcraft and Female Sexuality in *The Winter's Tale*', *A Journal for the Interdisciplinary Study of Literature*, 35.1 (March), 95–112.

Rosenthal, Earl (1971), 'Plus ultra, non plus ultra, and the Columnar Device of Emperor Charles V', *Journal of the Warburg and Courtauld Institute*, 34, 204–28.

Rossi, Paolo (1968), *Francis Bacon: From Magic to Science*, London: Routledge.

Rossi, Paolo (2001), *The Birth of Modern Science*, Oxford: Blackwell.

Rusche, Harry (1969), 'Edmund's Conception and Nativity in *King Lear*', *Shakespeare Quarterly*, 20.2 (Spring), 161–4.

Russell, Bertrand (2015), *In Praise of Idleness*, Philadelphia: The Great Library Collection by R. P. Pryne.

Sabra, Abdelhamid I. (trans. and ed.) (1989), *The Optics of Ibn al-Haytham, Bks I–III, On Direct Vision*, 2 vols, London: Warburg Institute.

Sabra, Abdelhamid I. (1994), *Optics, Astronomy and Logic: Studies in Arabic Science and Philosophy*, Yarmouth: Variorum.

Sabra, Abdelhamid I. (trans. and ed.) (2002), *Kitāb al-Manāẓir, Bks IV–V, On Reflection and Images Seen by Reflection*, 2 vols, Kuwait City: National Council for Culture, Arts and Letters.

Sabra, Abdelhamid I. (2007), 'The "Commentary" that Saved the Text: The Hazardous Journey of Ibn al-Haytham's Arabic *Optics*', *Early Science and Medicine*, 12, 117–33.

Sands, R. Kathleen (2004), *Demon Possession in Elizabethan England*, Westport, CT: Praeger.

Sawday, Jonathan (1995), *The Body Emblazoned: Dissection and the Human Body in Renaissance Culture*, New York: Routledge.

Sawday, Jonathan (2007), *Engines of the Imagination: Renaissance Culture and the Rise of the Machine*, New York: Routledge.

Schleiner, Winfried (1983), 'Orsino and Viola', *Shakespeare Studies*, 16, 135–41.

Schleiner, Winfried (1985), 'Imaginative Sources for Shakespeare's Puck', *Shakespeare Quarterly*, 36.1, 65–8.

Schoenfeldt, Michael (2010), *A Companion to Shakespeare's Sonnets*, Chichester: John Wiley & Sons Ltd.

Schrödinger, Erwin (1983), 'The Present Situation in Quantum Mechanics', in John A. Wheeler and Wojciech H. Zurek (eds), *Quantum Theory and Measurement*, Princeton: Princeton University Press, 152–67.

Scot, Reginald (1584), *The Discoverie of Witchcraft*, London: William Brome.

Scott, Hamish (ed.) (2015), *The Oxford Handbook of Early Modern European History, 1350–1750*, 2 vols, Oxford: Oxford University Press, vol. 1.

Seneca, Lucius Annaeus (1917–25), *Moral Epistles*, trans. Richard M. Gummere, Cambridge, MA: Harvard University Press, The Loeb Classical Library.

Seng, Peter J. (1967), *The Vocal Songs in the Plays of Shakespeare: A Critical History*, Cambridge, MA: Harvard University Press.

Shapin, Steven (1994), *A Social History of Truth: Civility and Science in Seventeenth-Century England*, Chicago: University of Chicago Press.

Shapin, Steven (1996), *The Scientific Revolution*, Chicago: University of Chicago Press.

Shapin, Steven (1998), 'The Early Modern Man of Science', in Katherine Park and Lorraine Daston (eds), *The Cambridge History of Science*, Cambridge: Cambridge University Press, 179–91.

Shapin, Steven, and Simon Schaffer (1985), *Leviathan and the Air-Pump: Hobbes, Boyle, and the Experimental Life*, Princeton: Princeton University Press.

Sheidley, William E. (1994), 'The Play(s) within the Film: Tom Stoppard's *Rosencrantz & Guildenstern Are Dead*', in Michael Skovmand (ed.), *Screen Shakespeare*, Aarhus: Aarhus University Press, 99–112.

Sherman, William H. (1995), *John Dee: The Politics of Reading and Writing in the English Renaissance*, Amherst: University of Massachusetts Press.

Sibly, Ebenezer (1784), *A New and Complete Illustration of Celestial Science*, 4 vols, London, vol. 1.

Sidney, Philip (1999) [1973], *The Countess of Pembroke's Arcadia: The Old Arcadia*, ed. Katherine Duncan-Jones, Oxford: Oxford University Press.

Simon, Gérard (2003), *Archéologie de la vision: l'optique, le corps, la peinture*, Paris: Seuil.

Simpson, James (2007), 'Diachronic History and the Shortcomings of Medieval Studies', in Gordon McMullan and David Matthews (eds), *Reading the Medieval in Early Modern England*, Cambridge: Cambridge University Press, 17–30.

Smith, A. Mark (1987), *Descartes's Theory of Light and Refraction: A Discourse on Method*, Philadelphia: American Philosophical Society.

Smith, A. Mark (2001), *Alhacen's Theory of Visual Perception, I: Introduction and Latin Text (Bks I–II–III)*, Philadelphia: Sources and Studies in the History and Philosophy of Classical Science.

Smith, Bruce R. (1999), *The Acoustic World of Early Modern England: Attending to the O-Factor*, Chicago: University of Chicago Press.

Smith, Bruce R. (2013), 'Within, Without, Withinwards: The Circulation of Sound in Shakespeare's Theatre', in Farah Karim-Cooper and Tiffany Stern (eds), *Shakespeare's Theatres and the Effects of Performance*, London: Bloomsbury, 171–94.

Smith, Pamela H., and Paula Findlen (eds) (2002), *Merchants and Marvels: Commerce, Science, and Art in Early Modern Europe*, London: Routledge.

Snow, C. P. (1993) [1961], *The Two Cultures and the Scientific Revolution: The Rede Lecture 1959*, Cambridge: Cambridge University Press.

Snow, Edward A. (1980), 'Sexual Anxiety and Male Order of Things in *Othello*', *English Literary Renaissance*, 10, 384–412.

Sokal, Alan (1996), 'Transgressing the Boundaries: Towards a Transformative Hermeneutics of Quantum Gravity', *Social Text*, 46/7 (Spring/Summer), 217–52.

Sokol, B. J. (2003), *A Brave New World of Knowledge: Shakespeare's* The Tempest *and Early Modern Epistemology*, London: Associated Press.

Spade, Paul Vincent (1999), *The Cambridge Companion to Ockham*, Cambridge: Cambridge University Press.

Spenser, Edmund (2013), *The Faerie Queene*, ed. A. C. Hamilton, revised 2nd edn, New York: Routledge.

Spiller, Elizabeth (2007), *Science, Reading, and Renaissance Literature: The Art of Making Knowledge, 1589–1670*, Cambridge: Cambridge University Press.

Spiller, Elizabeth (2009), 'Shakespeare and the Making of Early Modern Science: Resituating Prospero's Art', *South Central Review*, 26.1/2 (Winter/Spring), 24–41.

Stern, Tiffany (2015), 'Time for Shakespeare: Hourglasses, Sundials, Clocks, and Early Modern Theatre', *Journal of the British Academy*, 3, 1–33.

Sternfeld, F. W. (1963), *Music in Shakespearean Tragedy*, London: Routledge & Kegan Paul.

Stoppard, Tom (1967), *Rosencrantz and Guildenstern Are Dead*, London: Samuel French.

Stoppard, Tom (1972), *Jumpers*, London: Faber and Faber.

Stoppard, Tom (1999), *Hapgood*, in *Plays 5*, London: Faber and Faber.

Stoppard, Tom (2009), *Arcadia*, London: Faber and Faber.

Strohm, Paul (1998), *England's Empty Throne: Usurpation and the Language of Legitimation, 1399–1422*, New Haven: Yale University Press.

Stubbes, Philip (1882) [1583], *Anatomy of Abuses*, 2 vols, reprint, London: New Shakespeare Society.

Sullivan, Erin (2013), 'A Disease unto Death: Sadness in the Time of Shakespeare', in Helena Carrera (ed.), *Emotions and Health, 1200–1700*, Leiden: Brill, 159–83.

Symonds, Robert Wemyss (1947), *A History of English Clocks*, London: Penguin Books.

Szönyi, György E. (2004), *John Dee's Occultism: Magical Exaltation through Powerful Signs*, Albany: State University of New York Press.

Taylor, Gary (1995), 'Shakespeare and Others: The Authorship of *Henry the Sixth, Part One*', in Leeds Barroll (ed.), *Medieval and Renaissance Drama in England*, vol. 7, 145–203.

Taylor, John (1630), 'Iacke a Lent, His Beginning and Entertainment [. . .]', in *All the workes of Iohn Taylor the water-poet Beeing sixty and three in number. Collected into one volume*, London.

Taylor, Michael (ed.) (2003), *1 Henry VI*, The Oxford Shakespeare, Oxford: Oxford University Press.

Teillard, Ania (1948), *Le symbolisme des rêves*, Paris: Stock.

Thier, Jean du (trans.) (1566), *Les Louanges de LA FOLIE, Traicté fort plaisant en forme de Paradoxe*, Paris: Hertman Barbé.

Thomas Neely, Carol (2004), *Distracted Subjects: Madness and Gender in Shakespeare and Early Modern Culture*, Ithaca and London: Cornell University Press.

Thompson, E. P. (1967), 'Time, Work-Discipline, and Industrial Capitalism', *Past and Present*, 38.1, 56–97.

Thomson, Peter (1997) [1983], *Shakespeare's Theatre*, 2nd edn, New York: Routledge.

Tolomio, Ilario (1993), 'The "Historia Philosophica" in the Sixteenth and Seventeenth Centuries', in Giovanni Santinello et al. (eds), *Models of the History of Philosophy, Vol. I: From its Origins in the Renaissance to the 'Historia Philosophica'*, International Archives of the History of Ideas 135, Dordrecht: Kluwer Academic Publishers.

Tournon, André (1983), *Montaigne. La glose et l'essai*, Lyon: Presses Universitaires de Lyon.

Traister, Barbara Howard (1984), *Heavenly Necromancer: The Magician in English Renaissance Drama*, Columbia: University of Missouri Press.

Traub, Valerie (1992), *Desire and Anxiety: Circulations of Sexuality in Shakespearean Drama*, London and New York: Routledge.

Turner, Henry S. (2006), *The English Renaissance Stage: Geometry, Poetics, and the Practical Spatial Arts 1580–1630*, Oxford: Oxford University Press.

Usher, Peter D. (2010), *Shakespeare and the Dawn of Modern Science*, Amherst, NY: Cambria.

Vanden Heuvel, Michael (2001), '"Is Postmodernism?": Stoppard among/against the Postmoderns', in Katherine E. Kelly (ed.), *The Cambridge Companion to Tom Stoppard*, Cambridge: Cambridge University Press, 213–28.

Venet, Gisèle (2014), 'More "flies in amber"? or, of Topicalities, Dictionaries and Poetics in *Love's Labour's Lost*', *Études anglaises*, 67.3, 274–89.

Vickers, Brian (1968), *Francis Bacon and Renaissance Prose*, Cambridge: Cambridge University Press.

Vickers, Brian (ed.) (1984), *Occult and Scientific Mentalities in the Renaissance*, Cambridge: Cambridge University Press.

Vickers, Brian (1992), 'Francis Bacon and the Progress of Knowledge', *Journal of the History of Ideas*, 53.3, 495–518.

Vickers, Brian (2007), 'Incomplete Shakespeare: Or, Denying Coauthorship in *1 Henry VI*', *Shakespeare Quarterly*, 58.3 (Autumn), 311–52.

Wall, Wendy (2001), 'Why Does Puck Sweep? Fairylore, Merry Wives, and Social Struggle', *Shakespeare Quarterly*, 52.1 (Spring), 67–106.

Walsham, Alexandra (1999), *Providence in Early Modern England*, Oxford: Oxford University Press.

Walter, J. H. (ed.) (1954), *King Henry V*, Arden Second Series, London: Methuen.

Webster, Charles (1975), *The Great Instauration: Science, Medicine and Reform, 1626–1660*, London: Duckworth.

Webster, John (1996), *The White Devil*, in *The Duchess of Malfi and Other Plays*, ed. René Weis, Oxford: Oxford University Press, 1–101.

Weeks, Sophie (2007), 'Francis Bacon and the Art-Nature Distinction', *Ambix*, 54.2, 117–45.

Wells, Marion A. (2007), *The Secret Wound: Love-Melancholy and Early Modern Romance*, Stanford: Stanford University Press.

Wells, Robin Headlam (1994), *Elizabethan Mythologies: Studies in Poetry, Drama and Music*, Cambridge: Cambridge University Press.

Wells, Stanley, Gary Taylor, John Jowett and William Montgomery (eds) (2005), *The Oxford Shakespeare: The Complete Works*, 2nd edn, Oxford: Clarendon Press.

Whitlock, Keith (1999), 'Shakespeare's *The Tempest*: Some Thought Experiments', *Sederi*, 10, 167–84.

Wickham, Glynne (1975), 'Masque and Anti-Masque in *The Tempest*', *Essays and Studies*, 28, 1–14.

Williams, Wes (2012), '"L'humanité du tout perdue?" Early Modern Monsters, Cannibals and Human Souls', *History and Anthropology*, 23.2 (June), 235–56.

Wilson, Miranda (2014), *Poison's Dark Works in Renaissance England*, Plymouth: Bucknell University Press.

Wölfflin, Heinrich (1952) [1915], *Les Principes fondamentaux de l'histoire de l'art*, trans. C. and M. Raymond, Paris: Plon.

Wootton, David (2015), *The Invention of Science: A New History of the Scientific Revolution*, London: Penguin.

Worral, John (1998), *The Routledge Encyclopedia of Philosophy*, ed. E. Craig, 10 vols, London: Routledge, vol. 8.

Woudhuysen, H. R. (ed.) (1998), *Love's Labour's Lost*, The Arden Shakespeare, Walton-on-Thames: Thomas Nelson.

Wright, Thomas (1971) [1604], *The Passions of the Minde in Generall*, reprinted with an introduction by Thomas O. Sloan, Chicago and London: Urbana.

Yates, Frances A. (1936), *A Study of Love's Labour's Lost*, Cambridge: Cambridge University Press.

Yates, Frances A. (1964), *Giordano Bruno and the Hermetic Tradition*, London: Routledge & Kegan Paul.

Yates, Frances A. (1975), *The Last Plays*, London: Routledge & Kegan Paul.

Younge, Richard (1646), *The Cure of Misprision*, London.

Zaslavskii, Oleg B. (2005), 'The Little in a Non-Euclidean World: On the Artistic Space in Tom Stoppard's Film and Play "Rosencrantz and Guildenstern Are Dead"', *Sign Systems Studies*, 33.2, 343–67.

Index